D1325368
SYSTEM COPY

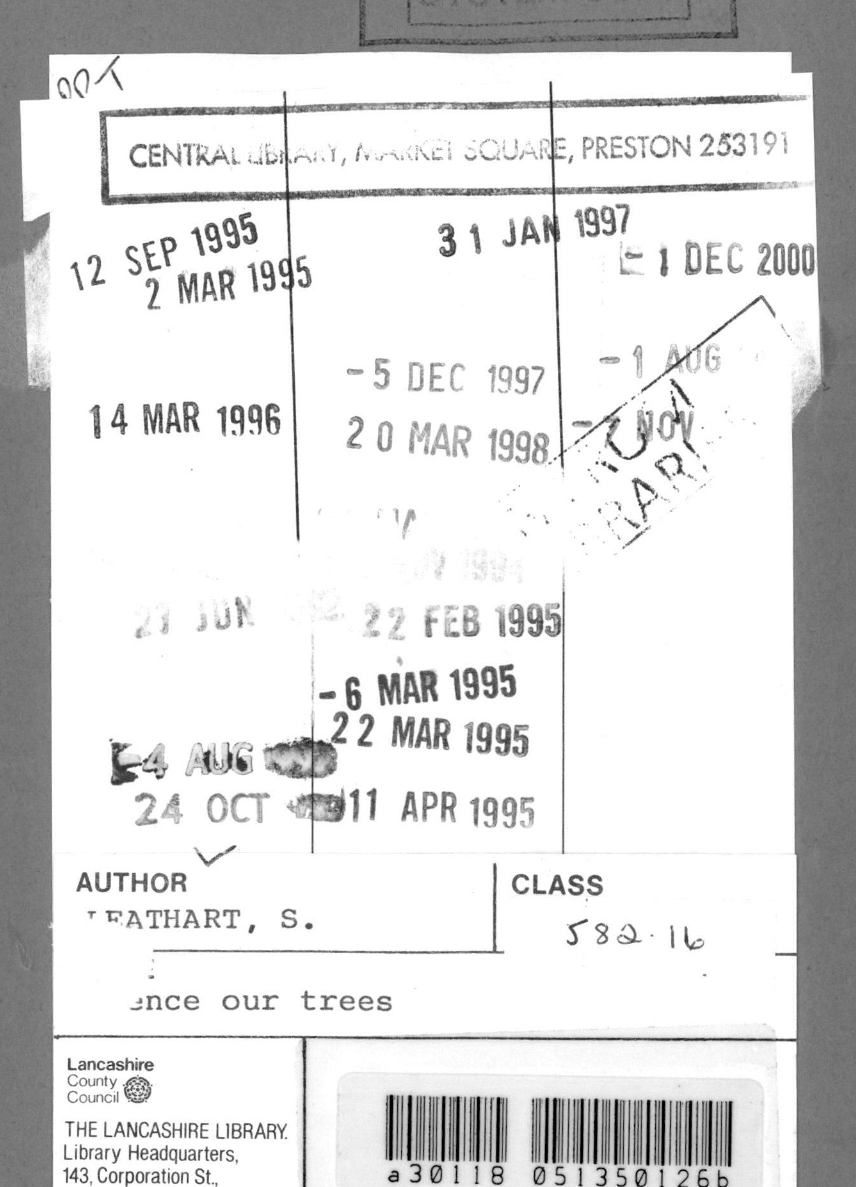
CENTRAL LIBRARY, MARKET SQUARE, PRESTON 253191
12 SEP 1995
2 MAR 1995
31 JAN 1997
1 DEC 2000
14 MAR 1996
5 DEC 1997
20 MAR 1998
22 FEB 1995
6 MAR 1995
22 MAR 1995
24 OCT
11 APR 1995
AUTHOR
LEATHART, S.
CLASS
582.16
ence our trees

WHENCE OUR TREES

To Peter,
who knows the worth of these trees

WHENCE OUR TREES

Scott Leathart M.B.E., M.A., F.L.S.

Foreword by
His Royal Highness, The Prince of Wales,
Duke of Cornwall

foulsham

LONDON • NEW YORK • TORONTO • SYDNEY

ACKNOWLEDGEMENTS

The publishers are grateful to the following organisations and individuals for permission to reproduce their photographs in this book.

John Beach: Golden Larch (facing p.97); Japanese Black Pine (facing p.144); Lodgepole Pine (facing p.145); Black Walnut (facing p.160); Cherry Birch (facing p.161); Pin Oak (facing p.176); Honey Locust (facing p.193); Pagoda Tree (facing p.208); Yellow Buckeye (facing p.225); Foxglove Tree (facing p.241); Indian Bean Tree (facing p.241).

Royal Botanic Gardens, Kew: The illustrations of the plant hunters between pages 24 and 25.

Royal Forestry Society of England, Wales and N. Ireland: Grand Fir (facing p.80); Morinda Spruce (facing p.97); Serbian Spruce (facing p.113); Western Hemlock (btm. rt., facing p.113); Corsican Pine (facing p.144); Lombardy Poplar (facing p.145).

The remaining photographs were supplied by the author and were taken by him during his visits to the countries concerned.

foulsham
Yeovil Road, Slough, Berkshire SL1 4JH

ISBN 0-572-01675-2

Printed in Great Britain at St Edmundsbury Press, Bury St Edmunds

Contents

Foreword

by

His Royal Highness, The Prince of Wales, Duke of Cornwall

When the Royal Forestry Society of England, Wales and Northern Ireland was founded one of its objectives was "to encourage the cultivation of trees. . . of recent introduction . . . as may be found suitable for the climate of the United Kingdom with a view to facilitating the study of their habitats, uses, value and cultural requirements". A century later the list of these exotics or non-indigenous species has been considerably extended. Many introduced varieties are part of the glory of our countryside for few would dispute the importance of such trees as the Wellingtonia (*Sequoiadendron giganteum*), the Noble Fir (*Abies procera*), the Common Walnut (*Juglans regia*) or the Horse Chestnut (*Aesculus hippocastanum*). Yet they have all come from other countries and feature in Scott Leathart's record of 104 of such examples. Many we have come to regard as our own, while others, such as the Sitka Spruce (*Picea sitchensis*), some of us might have wished, on purely aesthetic grounds, would have remained in their own homelands!

Scott Leathart, the Society's former Editor and Secretary, has produced a splendid and valuable work of reference and enjoyment, which not only describes the performance of the trees in Britain and the geographic location and description of outstanding specimens, but also their natural distribution and habitats. In addition he provides a most useful insight into the work of the plant explorers who contributed so greatly to the wealth of the British flora.

I am delighted, as President of the Royal Forestry Society to have this opportunity of welcoming a book which will become a standard reference work for foresters and gardeners.

Preface

The British Isles became detached from the European continental landmass relatively soon after the return of life to the land following the obliteration of most living things in the last great Ice Age, and the water barrier dividing the two areas of land put an end to the natural migration of the majority of plants and animals.

At the time of the severance, some 6,000 years ago, it seems from pollen evidence buried in peat and bogs, that only about 35 species of trees had been able to make the journey from ice-free areas in the south and spread over the land freed by the retreating ice.

It is with some of the 600 or so species and varieties which now grow in Britain that this book is principally concerned: when they were brought here, by whom and how they have fared since their assisted migration.

Some species, thought to have arrived naturally, may have been brought here by Neolithic people. Others were probably brought here by the Romans, notably the Walnut and the Sweet Chestnut. A trickle of introductions may well have been made by Crusaders and travellers in medieval times, but as seafaring prowess developed and sailors in ever bigger ships ranged further and further afield, the trickle became a flood and subsequently a spate until, by the end of the 19th century, few trees indigenous to the temperate regions of the Northern Hemisphere had not been brought to Britain and grown here, with varying degrees of success. But one, the Dawn Redwood, came as late as 1948 after escaping the attentions of plant hunters in China until 1941.

It is to the many intrepid explorers, to tireless and dedicated plant hunters, to observant officials sent by their governments to far-off places, to medical men travelling as ships' doctors and to men of learning and substance bitten by the travel bug that we owe our present extensive heritage; our thanks go to all these and, no doubt, to the many ordinary travellers and pilgrims who brought back home handfuls of nuts and seeds. Some of them are household words, others come to life again in the English or scientific names of the trees they first saw and described; and yet others suffer a public anonymity but are remembered with affection and respect by those concerned with the care and propagation of the trees which they introduced.

This work is not a textbook, neither is it an aid to identification. It is, rather, an attempt to present a picture, to those with a knowledge and love of trees, of some of the species brought to Britain and America from other countries, showing them in their natural habitat, describing how and when they reached their adopted homes and how they have fared since their arrival. It also details the uses to which the trees have been put, both commercial and aesthetic, since their establishment beyond their areas of natural distribution.

Much of the information on the introduction and the early performance of the trees has been gleaned from two monumental works, one, *Arboretum et Fruticetum Britannicum*, written by J.C. Loudon in 1838, and the other, *The Trees of Great Britain and Ireland* by H.J. Elwes and A.H. Henry published between 1906 and 1913. Subsequent data comes from other works listed in the bibliography and from my personal visits to Continental Europe, North America, India, Nepal, the Far East, South Africa, Japan, Australia and New Zealand where I have been able to see many of the trees described in their natural habitat. In some cases it has been possible to trace the history of individual trees from first introduction to the present day.

The recent measurements, so generously provided by Alan Mitchell from his unique records, and by the Tree Register of the British Isles, which incorporate them, attempt to give a geographical spread and thus something like average dimensions, rather than indicate merely the tallest and the largest. At least a proportion of the measured trees will be accessible to most readers (with the owners' permission) should they wish to go and see them.

The measurements are given as '36.3m × 4.5m (120ft × 15ft 3in)'. The first figure is the height of the tree, and the following figure is the girth (circumference) of the bole at a standard breast height of 1.5 metres (5ft 3in) unless otherwise stated. All the measurements, including those made by the late R.C.B. Gardner, a former Secretary of the Royal Forestry Society, were originally made in feet and inches and are more precise than the metric figures.

The chapter on some of the better-known plant explorers associates individual species with the men who discovered them, and who often endured considerable privations, even dangers, to bring them to us. Their stories, though briefly told, give extra meaning to both English and scientific names and an added interest to the trees concerned.

My thanks are due, first and foremost, to Alan Mitchell, formerly silviculturalist with the Forestry Commission who has been described by Hugh Johnson as 'the great tree connoisseur of our generation'. Without his tree measurements this book would have lost much of its interest. I am also deeply indebted to Brian Mulligan, Director of the University of Washington Arboretum, Seattle; to Dr Henry T. Skinner, formerly Director of the U.S. National Arboretum, Washington D.C.; and to Professor Clarence E. Lewis, formerly Professor of Landscape Horticulture at Michigan State University. All have given me valuable information.

I am grateful to the Librarian, The Commonwealth Forestry Institute, Oxford; and to the Director of the Royal Forestry Society of England, Wales and Northern Ireland, Tring, Herts., for permission to borrow books for protracted periods. I also thank the Editor, Quarterly Journal of Forestry, for permission to reproduce extracts from several issues.

Scott Leathart
October 1991

THE PLANT HUNTERS

Japan

Englebert Kaempfer (1651-1715)

One of the earliest European visitors to the then closed country of Japan, and one who was observant and interested in botany was Englebert Kaempfer, a German doctor. His name lives on in that of the Japanese Larch *Larix kaempferi*. He was sent out to Japan around 1689 as physician to the Governor of the Dutch East India Company's Deshima trading post and he was amazed to see so many new types of trees and other plants. His movements were severely restricted and he was not allowed to wander or collect plants at will. Kaempfer was made to take oaths, sealed in blood, that he would not fraternise with the Japanese people. He was permitted to make an annual pilgrimage to the Imperial Court between March and May in the company of over 150 officials. On these rare journeys, he carried with him a 'very large Javan box', which had a compass concealed beneath it. Into this he stuffed 'plants, flowers and branches of trees' plucked from the roadside, presumably with considerable agility since he was not allowed to dismount from his horse except when the path was too steep. At such times, he managed to evade his escort and stray from the roadway to fill his handkerchief with 'a great many of the most curious and scarcest plants'. But such comparative freedom never lasted long and by nightfall he was locked in his quarters after being treated like a circus animal; he was made to dance and sing, and jump about miming European manners and customs for the edification of the local people.

On his return home, Kaempfer practised as a physician but also found time to write, in 1712, his *Amoenitates Exotica*, the fifth part of which was devoted to plants and included the first-ever description of the Japanese Larch, the Maidenhair Tree *Ginkgo biloba* as well as several *Prunus* and *Magnolia* species.

Carl Peter Thunberg (1775-1829)

There must have been an interegnum when no plant hunting was done, for it was not until 85 years after Kaempfer went home that the Swedish doctor, Carl Peter Thunberg, arrived on the scene. He went out to Japan assigned to the same post as Kaempfer. Although trading restrictions were even more severe, social barriers had to some extent come down. Thunberg was allowed to make botanical excursions whenever the spirit moved him, albeit escorted by 'head and sub-interpreters, head and sub-banjoses, purveyors and a number of servants'. And although, unlike Kaempfer, he was never treated like some sort of performing animal, he was often beseiged by local astrologers and by physicians seeking advice about their patients whom he was never permitted to see. He made a great many Japanese friends who subsequently corresponded with him and sent him seeds. After he returned home he wrote of his finds in his

Flora Japonica, – a work of more substance than Kaempfer's, published in 1784. Amongst the plants he mentioned was his permanent memorial, the Japanese Black Pine *Pinus thunbergii* and the Hiba *Thuyopsis dolobrata*. He succeeded Linnaeus as professor at Uppsala University in this same year.

Philipp Franz von Siebold (1790-1866)

The international situation was very different 50 years later when a German, Philipp Franz von Siebold, a German went out to occupy the same post as Kaempfer and Thunberg. Holland had temporarily lost her Far East possessions as a result of the Napoleonic Wars and Deshima was the only place where the Dutch flag still flew. Siebold was a skilled doctor and of great value to the local people being an eye specialist as well as one who could perform cataract operations and enable the blind to see. He was thus able to exert enormous influence, and although on his arrival he found all the old restrictions, he was soon permitted to travel far and wide on the pretext of visiting his patients. Somehow he managed to obtain maps (the possession of which was a treasonable offence amongst the Japanese) and he contrived to send a consignment of plants home in 1829, taking a further 485 specimens with him when he left Japan in 1830. His *Flora Japonica*, published in 1842, was illustrated by Japanese artists, something of a scoop at the time.

Siebold returned to Japan in 1859 to become confidential advisor to a Japanese privy councillor and to introduce European sciences. Unfortunately, he meddled too much in politics and was forced to leave in 1862. Despite his preoccupations with political intrigue and, no doubt, lucrative forays into medicine, having a somewhat mercenary turn of mind, he did manage more botanising, bringing home with him the Japanese Crabapple *Malus floribunda* and two other trees which are his memorial – *Prunus sieboldii* and the South Japanese Hemlock *Tsuga sieboldii*.

John Gould Veitch (1839-1870)

At about this time it was to be an Englishman's turn to collect plants in Japan. He was John Gould Veitch, grandson of the founder of the famous Exeter nursery. He went out to Japan in 1860, undaunted by shipwreck off Ceylon en route. He settled in Nagasaki but at first was not permitted to range beyond a 16 kilometre (10 mile) radius of the city. Nevertheless, he collected 50 plants and potted them up in his garden. He also filled a box with 70 different kinds of seeds. He took the trouble to learn Japanese, perhaps from the ladies, for he rated them 'friendly and not shy' but not in the least impressed or attracted by European whiskers, of which Veitch had an ample supply.

After a while he managed to obtain permission to go to Yokohama where he met the British consul, Rutherford Alcock, who was journeying to the sacred mountain Fujiyama, where no European had been before. Veitch joined the party and, to make it official, was appointed 'Botanist to Her Britannic Majesty's Legation at Yedo' which made the young man more than usually proud. On the mountain Veitch discovered his own fir *Abies veitchii* as well as a new spruce, Alcock's Spruce *Picea bicolor*, which he named after his companion.

For most of his remaining stay in Japan, Veitch lived with Alcock in Tokyo where his movements were very restricted, but he was able to send out native collectors armed with drawings of what to look for, and by the time he left for home at the end of the year he had amassed a considerable collection. This included no less than 17 conifers, among them the Japanese Larch which Kaempfer had discovered, but not, apparently, his own fir, which was not introduced until 1879 when Maries sent seeds to Veitch's nursery. Two magnolias, *M. stellata* and *M. soulangiana* were also in the consignment.

Veitch made another trip, this time to Australia and the South Sea Islands, but on his return became ill and died of consumption at the early age of 31. Thus a promising career of a friendly, gentle and determined young man was tragically cut short.

Charles Maries (1852-1901)

The Japan to which Charles Maries was to go some 17 years later was vastly different from that of Kaempfer's day. There was an industrial exhibition in the Mikado's palace which was only entered on pain of death by the common people in former days. Maries now had no difficulty in travelling wherever he wished. He had been a foreman at Veitch's nurseries and at an early age had been bitten by the travel bug. His visit to Japan, with occasional trips to China, lasted three years. He arrived there in April 1877 and, after exploring the Inland Sea area and the environs of Kobe, he visited Kyoto, Yokohama and Tokyo but was not impressed by the neglected Japanese gardens which he saw – a neglect which seemed to stem from the generally more relaxed atmosphere in the country.

His first long journey in Japan took him overland (ships were not sailing because of the civil war in the South) up the length of Honshu to Aomori and thence across the straits to Hokkaido where, on Mt Hakkoda, surrounded by an almost impenetrable mass of bamboo (containing two black bears) he found his own fir *Abies mariesii*. There he also collected the seed of Veitch's Fir which Veitch had been unable to do. He returned to his base to find it burnt to the ground. His baggage had been rescued, only to suffer later when the ship he took back to Honshu sprang a leak. The cargo of seaweed expanded and the captain had to beach the craft. The baggage was put aboard a small boat which promptly capsized so Maries had to re-collect those specimens which had been ruined.

After an excursion up the Yangtse, where the Chinese, whom he distrusted as much as they did him, looted his seemingly ill-fated baggage, he returned home in 1880 bringing with him more than 500 plants. In 1882 he went out to India, where, on the recommendation of Sir Joseph Hooker (Director of the Royal Botanic Gardens, Kew, at that time),he became Superintendent of the Maharajah of Gwalior's gardens until his death 20 years later.

China

James Cunningham (d. c.1709)

Few details are known about the personal life of James Cunningham who was a Scot and one of the first amateur botanists to visit China. He sailed there in 1698 as a surgeon with the English East India Company. He stayed in Amoy for a while and then managed to procure paintings of 'near 800 plants in their natural colours', as well as collecting dried plants, shells, insects etc. He returned to England in 1699 but by 1702 was back in China again, this time in Chusan, where he discovered the Japanese Cedar *Cryptomeria japonica* (later to be found in Japan by Kaempfer who gave it its rather misleading name) as well as the tree which is his memorial, the Chinese Fir *Cunninghamia lanceolata*.

Cunningham appears to have remained in the Far East until 1709, with revolutions, massacres and lootings dogging his steps in both the Maccassars and Cochin China. He was last heard of in a letter sent from Calcutta in January 1709, he presumably died on the voyage home.

William Kerr (d.1815)

The first professional collector to go to China was William Kerr, sent out there in 1803 by Sir Joseph Banks who was the first Director of the Royal Botanic Gardens at Kew. He stayed there for eight years but never managed to range further afield in China than Canton and Macao; he did though, make an expedition to the Philippines and Cochin China, losing most of his 700 plants in a storm on his way back to Macao. He sent hundreds of plants home over the years, in special cases attended by a Chinese gardener. He supervised the drawing of plants by native artists, as well as maintaining a garden in Macao where he attempted, without much success, to establish European plants sent out from Kew. Towards the end of his long stay in the East his energy appears to have lessened, and it seems likely that he took to opium after 'associating with inferior persons'. Nevertheless, Banks rewarded him by appointing him Superintendent of the new botanic gardens in Ceylon; but he died in 1815, only 15 months after taking over. Among his introductions was the Chinese Juniper *Juniperus chinensis*.

Robert Fortune

After the Opium Wars, with the ceding of Hong Kong to the British, the temporary occupation of Chusan and the opening up of four new Treaty

Ports, China became much more accessible to Europeans and the Horticultural Society decided to send out a collector. They chose Robert Fortune, a foreman at their Chiswick garden. He arrived in Hong Kong in July 1843 and made his base in the new colony from where he made forays into China proper, travelling as far north as Shanghai where he was often surprised to find snow on his bedroom floor. Despite an edict forbidding Europeans to travel further than 48 kilometres (30 miles) inland from a Treaty Port, Fortune managed to reach Soochow, by dint of wearing a Chinese costume, and complete with a shaved head and pigtail. Needless to say, he was robbed and once had all his clothes stolen, but he remained quite undaunted, searching diligently for plants. He returned home in 1846 with 250 specimens in addition to those he had already sent ahead, only 35 of them dying en route. During this stay at home he wrote his first book, *Three Years Wandering in the Northern Provinces of China*, which contained two chapters on the cultivation of tea. This led to his being sent back to China, this time by the East India Company, with the mission to obtain tea seeds and plants for setting up tea gardens in India. He successfully completed this mission despite many vicissitudes, travelling further into China than any other European had ever done before. His plants were used to start tea gardens in Kumaon and Garhwal before he returned home in 1852.

So successful had he been that he was once again sent to China in search of even better tea plants. It was during this third trip that he discovered the Golden Larch *Pseudolarix amabilis*. On a previous expedition he had found the Chinese Cow's Tail Pine *Cephalotaxus fortuni*, which is one of his memorials, and the Chinese Weeping Cypress *Cupressus funebris*.

He returned home once more in late 1856 but was off again to China in 1868 in quest of tea, this time for the American government who wanted to experiment with it in the Southern States. Travelling extensively, he revisited old haunts, but there are few details of what he found. In 1860 he went to Japan, where he met Siebold and Veitch. He travelled as much as possible collecting seed, mainly from indigenous trees, settling for the final fortnight in Alcock's legation in Tokyo. From there both he and Veitch took passage home on the S.S. England with their collections, so that 'the whole poop was lined with glass cases crammed full of the natural products of Japan'. It is remarkable that neither man ever made any comment about the other, considering their identity of interests.

Augustine Henry (1857-1930)

For the rest of the century China was to remain a rich hunting ground for plants. Soon after qualifying as a doctor at Belfast University, Augustine Henry joined the Chinese Customs Service as a medical officer and Assistant Superintendent of Customs. After a few months in Shanghai, he was posted to Ichang at the extreme limit of navigation on the Yangtse River. This put him in a most advantageous position for exploring the flora of the country, an interest he developed from studying plants used as drugs by the Chinese. He had struck up a friendship with the Director at Kew, William Thistleton-Dyer, and he was soon sending regular shipments of plants home – the first consignment being in 1885, consisting of over 1,000 specimens.

Over the years he was posted to widely separated places – Hupeh, Szechwan, Yunan, Hainan and Taiwan – and wherever he was he ranged over the countryside in search of plants. These travels made him realise how quickly China was becoming deforested and he asked Kew to send out a man expressly to collect plants before they disappeared. But Kew was unable to oblige and Veitch's Nurseries agreed to find a man who was to be supported by funds supplied by Professor C.P. Sargent of the Arnold Arboretum, Boston, Mass.

On his return to England, Henry co-operated with H.J. Elwes in writing that monumental work, *Trees of Great Britain and Ireland*, after becoming Reader in Forestry at Cambridge University and later Professor of Forestry at Trinity College, Dublin, a post he held for 13 years.

Ernest Henry Wilson (1876-1930)

The man whom Veitch's Nurseries chose to go out to China was Ernest Henry Wilson. He had left school at 16 and by the time he was 21 was on the staff at the Royal Botanic Gardens, Kew and a winner of the Queen's Prize for Botany. He left for China in 1899 via the USA where he spent a week at the Arnold Arboretum and struck up firm friendship with his financial backer, Professor Sargent; a friendship which was to stand him in good stead for the rest of his life. He met Henry, who was in his last year of service in China and who must have given him invaluable advice. Fortunately he did not heed Veitch's general advice which was 'stick to one thing which you are after and do not spend time and money wandering about. Probably almost every worthwhile plant in China has now been introduced into Europe'.

The tree which Wilson wanted to find on this first expedition was the Dove-tree *Davidia involucrata*. Henry had found one which he located on a sketch-map but Wilson discovered that it had been felled to build a house. No other specimens were known except one over 1,600 kilometres (1,000 miles) away. Luckily Wilson came across one very much nearer and later located a grove of some 20 trees. These he watched until they produced seed so was able to collect a good number. He never again found a *Davidia* bearing seed.

Often on his expeditions Wilson followed in Henry's footsteps. He found the same mountainous regions of north west Hupeh particularly rich in trees and shrubs, all of which subsequently proved reliable in cultivation. Wilson returned home in 1902, bearing seed of 305 species and 900 herbarium specimens. Apart from the *Davidia*, the Paperbark Maple *Acer griseum* and Wilson's Spruce *Picea wilsoni* were among the trees in this introduction. All the specimens were established in Veitch's Coombe Wood Nurseries.

Wilson made three more expeditions to China, at Sargent's instigation, the second and third after he had emigrated to the USA and had become Keeper of Botany at the Arnold Arboretum. His last foray to China was in 1910, this time in search of a lily, *Lilium regale* which he eventually found near one of the trade routes into Tibet. Both he and his wife were killed in a car crash in America in 1930.

George Forrest (1873-1932)

Perhaps the last collector to be employed by a private patron was George Forrest. He had been something of a rolling stone, starting life as a pharmaceutical chemist and later roaming round Australia before returning to his native Scotland to work in the herbarium at the Royal Botanic Garden, Edinburgh. There it was that he was found by his future patron, Arthur Bulley who was a rich Liverpool cotton merchant, and later the founder of Bees Seeds. Bulley's previous experience of plant collecting by proxy through missionaries had been a dismal failure. His wife wrote in the Gardeners' Chronicle that the results had been 'the best international collection of dandelions to be seen anywhere'. Things were to change dramatically once the new collector had set foot in China.

Forrest arrived in Yunan, via Burma, in 1904 after escaping with his life when his party was set upon and mostly massacred by Batang Lamas. He was to spend 28 years collecting in China, mainly in the high mountain areas where rhododendrons (309 species) and primulas (67 species), his chief interests, were mostly to be found. The high altitude variety of Delavay's Fir *Abies delavayi* var. *forrestii* is a tree which remains a permanent memorial to this remarkably energetic man who sent home no less than 31,000 sheets of dried plants and an equal number of seed packets. Typical of him was his end. Although far from well, he had, during the day, personally packed up and loaded onto two mules seeds of some 500 plant species, when, taking his gun he set off to bag something for the pot. Reaching for a high bird, it fell dead, and so did he at the age of 58. He was buried at Teng-yueh in the country he had come to know so well.

The Himalayas

Nathaniel Wallich (1786-1854)

Born in Copenhagen Nathaniel Wallich went to India in 1807 as surgeon to the Dutch Settlement at Serampore. After the Napoleonic Wars, when the settlement was ceded to Britain, Wallich joined the East India Company as an army surgeon. In 1807 he was given temporary charge of the Botanic Gardens in Calcutta, becoming Director two years later, a post he held for 30 years. During this time he engaged native collectors, sending them far and wide into the Himalayas, and particularly into Nepal, a closed country in those days from which he was able to send many consignments of seeds to England. He went to Kathmandu in 1820 but, like all Europeans up to as recently as 1950, he was not permitted to range beyond the valley. However, he was able to persuade pilgrims and travellers to collect seed for him, and engaged a native artist to draw the specimens that they had brought back. These included the Tree Rhododendron *R. arboreum* and the Bhutan Pine *Pine wallichiana*, by which we remember him.

The increasingly anglicised Wallich came to England in 1828 and there wrote his *Plantae Asiaticae Rariores*, before returning to India again in 1833 making occasional forays into South East Asia. In 1847, broken in health, he returned to England where he died seven years later.

William Griffiths (1810-1845)

Wallich was succeeded by William Griffiths at the Calcutta Botanic Gardens, and he accompanied him on an expedition to Assam to investigate the Assamese tea plant and the climatic conditions under which it grew. Griffiths had entered the service of the East India Company as an assistant surgeon, but he was also a brilliant botanist. During the expedition he fell out with Wallich, mainly in the clash of personalities inevitable when a brash young man lacks tact in dealing with an experienced veteran. They also differed in opinion on the native tea plant which Griffiths reckoned should be used to stock the Assam gardens whilst Wallich recommended the Chinese plants. Griffiths was eventually proved right after 20 years of unsuccessful Chinese plantings.

On this and subsequent expeditions to Bhutan, Griffiths was also plant hunting. He too found the Bhutan Pine, and it is somewhat ironic that this tree, originally named *Pinus excelsa* and later *Pinus griffithii*, should end up as *Pinus wallichiana*; a circumstance unlikely to amuse Griffiths. However, one tree, the Sikkim Larch *Larix griffithiana*, still commemorates him. He died young after a short spell in charge of the gardens in Calcutta where his interest in the morphology of plants rather than their cultivation left the collection in a sorry state, much to the distress of Wallich when he returned there in 1844.

Joseph Dalton Hooker (1817-1911)

After graduating in medicine at Glasgow University, Joseph Dalton Hooker accompanied Captain Ross's expedition to the Antarctic as assistant surgeon aboard HMS Erebus. On his return he was sent to India by the Royal Botanic Gardens, Kew, arriving there in 1847. His mission was to compare the Antarctic flora with that of high altitudes in the tropics. Consequently, his three journeys during the four years he spent in India were confined to places high up in the Himalayas, mostly in Nepal where the Maharajah welcomed him, and in Sikkim where the Dewan (Prime Minister) opposed him, causing endless difficulties, even dangers. But with his friend, Dr Archibald Campbell, political agent at Darjeeling, (whose name was given to one of Hooker's finds, *Magnolia campbellii*) he managed to travel extensively in Sikkim and even reached some of the passes into Tibet. His main botanical interest was rhododendrons of which he found 43 species and collected the seed of most of them. His courage and determination were remarkable in the face of harassment by locals, starvation and an often harsh climate, as well as a terrain requiring great exertions to traverse it. Yet he lived to be 94 after serving 20 years as Director at Kew and giving his name to a large number of Himalayan plants.

North West America

David Douglas (1799-1834)

The son of a Perthshire stonemason, David Douglas became a gardener's boy at the age of ten. Self-educated, he obtained a post at the Glasgow Botanic Gardens at the age of 20 and was selected by the Horticultural Society to act as their collector in North America. His first visit in 1823 was concerned mainly with nurseries and gardens in New York and Philadelphia, with a small amount of collecting in the Eastern States. He returned home in 1824. The Society was so well pleased with his efforts that, in the following year, with additional assistance from the Hudson's Bay Company, he was sent to north west America.

It was eight months before the ship, the *Mary Anne*, arrived at the mouth of the Columbia River, and even then storms prevented it from entering the river for nearly another month. Eventually Douglas reached Fort Vancouver, the company's post, and spent the summer exploring the lower reaches of the Columbia, with two trips into the Cascade Mountains. He returned to the Fort with 500 specimens of plants as well as skins of birds and mammals which he sent home on the returning *Mary Anne*. Then, despite a knee injury, he made a month's trip up north to the Chehalis River before Christmas. Thus, in his first year he travelled over 3,000 kilometres (2,000 miles) on explorations. In 1826 he was to double that distance.

In March of that year he set out up the Spokane River to Kettle Fall and Okanagan, returning to the Fort in August, only to set out again, this time south along the valley of the River Williamette and on to the Umptqua River and the mountains to the south of it. Here he came across the tree he was so anxious to find – the Sugar Pine *Pinus lambertiana*, named after A.B. Lambert, secretary of the Horticultural Society at the time. The story of this discovery, the largest pine in the world, as told in Douglas' diary, is surely one of the most famous descriptions of the finding of any plant:

'At midday I reached my long-wished Pinus . . . I now state the largest one I could find that was blown down by the wind; three feet from the ground, 57 feet 5 inches in circumference; 134 feet from the ground 17 feet 5 inches; extreme length, 215 feet . . .'

The only way Douglas could collect any of the cones was by shooting them down, but this also attracted some of the local Indians armed with bows, arrows and knives. There followed a somewhat tense encounter which ended with Douglas sending them on a search for cones in return for some tobacco. He later wrote:

'They went, and as soon as out of sight I picked up my three cones and a few twigs, and made a quick retreat to my camp . . .'

Douglas was back at the Fort by the end of November 1826 and in March of the following year he set off on his longest journey which took him into Canada, over the Rockies to Edmonton and Winnipeg, and eventually to York Factory on the shores of Hudson's Bay where he took ship to England. He arrived there in triumph on 11 October 1827; for no collector before him had ever introduced so many hardy, ornamental plants – 210 in all. These included no less than 12 conifers, among them the Douglas Fir *Pseudotsuga menziesii*, his all-time memorial, and the Sitka Spruce *Picea sitchensis* which is now the mainstay of British upland forestry. This success in finding conifers prompted Douglas to write to Hooker 'You will begin to think that I manufacture pines at my pleasure'.

For two years he enjoyed the fame and the limelight which were his due, but in October 1829 he again sailed for America, arriving off the Columbia River in June 1830 to be prevented from travelling inland by tribal warfare amongst the Indians. Eventually he made a two-month trip up the Snake River, after which he sailed down the coast of California and Mexico. Part of the time down south he was accompanied by Coulter after whom he named their joint discovery, the Big-cone Pine *Pinus coulteri*. In August 1831 he sailed for the Sandwich Islands from where he sent home 670 more plant specimens. He eventually arrived back on the Columbia River in October.

There followed more excursions, some for surveying purposes such as those in the Mount Hood area and in Puget Sound, but all the time he was planning a mammoth journey to Alaska. He set out in March 1833 but, inexplicably, turned back when no more than halfway there. In June his canoe was dashed to pieces on the Fraser River losing his personal possessions as well as all his notes and specimens. Before the accident he had spun round in a whirlpool for an hour which must have been a terrifying experience.

After this he had almost had enough, but not quite. He made another excursion into the Blue Mountains, and attempted to climb Mount Hood before sailing once again to the Sandwich Islands (now Hawaii) on 4 November 1833, where he arrived in time for Christmas. The next six months were spent in climbing two volcanoes and travelling round the several islands, but on 14 July this courageous and indomitable explorer, who had endured hardship and escaped death on so many occasions, met a horrible end. He fell into a pit dug to trap the feral cattle and was gored to death by a beast which had fallen in before him. What other discoveries he might have made had he lived we shall never know, but he had shown what a treasure-house of plants north west America was, and he was to be followed by others inspired by his example and his success. The tree species which Douglas introduced, in addition to those mentioned above, were: Beautiful Fir *Abies amabilis*, Grand Fir *Abies grandis*, Noble Fir *Abies procera*, Vine Maple *Acer circinatum*, Big-leaf Maple *Acer macrophyllum*, Madrone *Arbutus menziesii*, Western White Pine *Pinus monticola*, Monterey Pine *Pinus radiata*, and Digger Pine *Pinus sabiniana*. He introduced more trees of commercial forestry use than any other man.

Theodore Hartweg (1812-1871)

Descended from a German family of gardeners, Theodore Hartweg worked for a time at the Jardine des Plantes in Paris before becoming a clerk at the Horticultural Society's gardens in Chiswick. He was sent out by the Society to California and Mexico in 1836 where, like so many early plant collectors, he mixed his searches with alarms and real dangers caused by the general unrest of this turbulent period in California's history. One of his first allotted tasks was to obtain seeds of the Golden Chestnut *Chrysolepis chrysophylla*, a tree which Douglas had discovered but had not introduced. In the course of a decade in these parts, Hartweg discovered the Monterey Cypress *Cupressus macrocarpa* and that most beautiful of pines, the Montezuma Pine *Pinus montezumae*, the hardier upland variety of which, *hartwegii*, is his memorial. He has also been credited with the introduction of the Redwood *Sequoia sempervirens*, but as the date, 1834, generally considered to be that of its introduction via St Petersberg, seems to fit with the date of early plantings in England, it is unlikely that Hartweg was the first to bring it here.

John Jeffrey (1826-c.1854)

So famous had the western states of America become for new conifers that, in 1850, a number of wealthy landowners in Britain banded together to form the Oregon Association. The object of the Association was to employ a collector who would go out there to gather seeds of all the trees which Douglas had discovered, and perhaps find some new ones. The Association chose John Jeffrey who was working at the Royal Botanic Garden, Edinburgh. He sailed in August 1850 on board the Hudson's Bay ship, *The Princess of Wales*, to York Factory in the Bay. In January 1851 he set out on his long, cold journey of 1,900 kilometres (1,200 miles) travelling mostly on snowshoes with dogs to drag his kit. He crossed the Rockies in April to start his first collecting season in what is now the British Columbia/Washington State border area. Here he discovered the Subalpine Whitebark Pine *Pinus albicaulis*, and on an early spring foray to Vancouver Island in 1852, he came across the Western

John Gould Veitch

Charles Maries

Robert Fortune

Augustine Henry

Ernest Henry Wilson

George Forrest

Nathaniel Wallich

William Griffiths

Joseph Dalton Hooker

David Douglas

Henry John Elwes

Hemlock *Tsuga heterophylla*. Then he moved south, following Douglas' footsteps, and discovered the Foxtail Pine *Pinus balfouriana* which he named after one of his sponsors in the Oregon Association. Further south, in the Sierra Nevada, he found the rare Macnab's Cypress *Cupressus macnabiana*, as well as his own pine *Pinus jeffreyi*, then considered to be a variety of *P. ponderosa* but now a separate memorial to him.

Despite these successes, Jeffrey did not live up to his early promise as an industrious collector. The further south he progressed the less became the material he collected until, after January 1854, consignments ceased altogether. Searches by local people and by members of the Oregon Association sent out to find him proved fruitless. Some say he died of thirst in Colorado, others suggest that he was murdered by renegade Spaniards for his mules, and yet others believe that he succumbed to the lure of the gold rush and met his fate with many others similarly hooked. At all events he was never seen or heard of again.

William Lobb (1809-1863)

It was not only the Oregon Association which sought to cash in on Douglas' discoveries. The famous nursery firm of Veitch had an eye to introducing the recently found trees to commerce, and they sent out William Lobb to collect for them.

First of all he went to South America from where, during 1847/48, he sent home the Roble Beech *Nothofagus obliqua* and the Patagonian Cypress *Fitzroya cupressoides* (named after Captain Fitzroy who commanded HMS Beagle, the ship in which Charles Darwin sailed to the Galapagos Islands) and that other strange conifer Prince Albert's Yew *Saxegothia conspicua*. In the summer of 1849 Lobb moved north to California and thence, by degrees, up to Oregon and the Columbia River. He had collected seeds of the Santa Lucia Fir *Abies bracteata*, previously discovered by Jeffrey, and both discovered and introduced the Western Red Cedar *Thuya plicata*. Although he did not actually introduce the Wellingtonia *Sequoiadendron giganteum* (this was done a month or two before Lobb by J.D. Matthews) it was from his living and dried specimens, collected in the Calaveros Grove, that the tree was first described and named by Dr Lindley after the great hero of the time, the Duke of Wellington, who died in 1853, the year when Lobb made a short visit home. He returned to settle in California the following year, and continued to collect and send consignments home until his death in San Francisco in 1863.

Henry John Elwes (1846-1922)

A wealthy Gloucestershire landowner, Henry John Elwes was also a man of great distinction. During his early years as a student in both Brussels and Dresden, and five years service as an officer in the Scots Guards, his thirst for travel grew greater as his interest in natural history expanded. He became a distinguished ornithologist, writing papers for *Ibis*. He was also a renowned entomologist, amassing the most complete collection of northern temperate butterflies ever known, the best 15,000 specimens of which he gave to the Natural History Museum. His travels all over the world made him one of the keenest observers of plants, especially lilies and trees.

Elwes was not perhaps a great collector, like others described here, but he put the seal on their discoveries by travelling far and wide to see for himself the habitats in which the trees grew and the other plant species which shared that habitat. He travelled over Europe, Asia Minor, North and South America, Japan, China and the Himalayas, observing with a keenness which few could match; all with a view to writing a book which would 'give a complete account of all the trees which grow naturally or are cultivated in Great Britain and which have attained or seem likely to attain a size which justifies their being looked on as timber trees'. In conjunction with Augustine Henry, he wrote *The Trees of Great Britain and Ireland*, the first volume appearing in 1906. It was to be followed by four more but so great was the amount of information amassed and the number of trees discovered while the work was in preparation, that it ended up as seven volumes the last of which was published in 1913. The work includes 1,933 pages, some 400 photographs of superb clarity and descriptions of 565 tree species. Its authenticity is impeccable, for the authors were as widely travelled as they were knowledgeable.

THE TREES

Ginkgo or Maidenhair Tree

Ginkgo biloba

This Chinese tree is in truth a living fossil. Trees identical to it were growing over much of the northern hemisphere 125 million years ago. It is the sole survivor of a great race of plants which once dominated the earth. It is a tree of great distinction with its spreading, rigidly ramified branches, curiously fan-shaped bright green leaves and grey-brown, deeply-fluted bowl which has strange protruberances in extreme old age. As the late Sir Albert Seward wrote in 1938, 'It appeals to the historic soul: we see it as an emblem of changelessness, a heritage from worlds too remote for our human intelligence to grasp, a tree which has in its keeping the secrets of an immeasurable past.' (See illustration facing p.32).

Distribution and habitat

Although opinions differ about the natural distribution of the Ginkgo, it is fairly certain that it originally grew wild in the mountain forests on the borders of Anwei and Chekiang in south east China. It is in this area that several of that country's curious conifers also seem to have originated, such as *Keteleeria davidiana, Pseudopanex amabilis* and *Torreya grandis*. It may well be native elsewhere, and is still to be found in one of its original growing areas, the hilly country of the Lower Yangtse Valley, where it flourishes in company with such trees as Golden Larch *Pseudolarix amabilis*, Chinese Sweet Gum *Liquidambar formosana* and various species of *Acer* and *Quercus*.

Over the centuries, cultivation has distributed the Ginkgo throughout most of central and southern China, and into Japan.

Despite the presence in Japan of some Ginkgoes reputed to be over 1,000 years old, it is certain that these trees are not indigenous to that country but were taken there from Chekiang, probably during the Sung Dynasty in the 10th century.

The Ginkgo is widespread in both China and Japan as an ornamental, much revered for its ancient associations.

Life history

The male and female flowers, born on trees of separate sexes, open in the spring before the leaves appear, the males in pendulous yellow catkins and the females like little, long-stalked acorns borne in ones and twos at the end of the spur shoots. These develop into orange-yellow plum-like fruits about 2.5cm (1in) in diameter, with an offensive-smelling outer fleshy layer which rots away to reveal a white, silvery nut that is edible and relished in China.

Growth of seedlings, although rapid at first, is not regularly maintained; height growth may cease for a whole season for no apparent reason.

Old trees reach great sizes. A tree near Tokyo temple, measured in 1884, was 31.7m (104ft) tall and girthed 7.3m (24ft); Elwes saw one in Kyoto which, although no great height, girthed 4.5m (15ft) and had a crown spread 90 paces in circumference.

Timber

The timber is soft, light brown in colour and has a silky sheen, but few of these trees are felled, and such timber as becomes available is used for chess-men, abacus beads and other trinkets for personal use.

Introduction, performance and use as an exotic

The Ginkgo was first seen by a European in 1690, by Kaempfer in Japan, and originally described by him in 1712. The first tree planted in Europe was at Utrecht, in Holland, about 1730. There is an old specimen there now which is possibly the same one. The first recorded Ginkgo in Britain was growing at the Mile End nursery of James Gordon in 1754, and was presumably planted a few years before that. Gordon must have propagated others from this original, for he sent one to Linnaeus in 1771, and the tree growing at Kew is probably another of them. This tree, a male planted in 1772, was 17m × 2.7m (56ft × 9ft) in 1888, and in 1904, Elwes found it to be 18.9m × 3.1m (62ft × 10ft 4in). By 1989, according to Mitchell, it had grown to 24.4m × 4.2m (80ft × 14ft 4in), and is still a fine tree in perfect health. Another specimen, thought to be the same age as the Kew tree, at Blaize Castle, near Bristol, was 20.7m × 2.7m (68ft × 9ft 3in) when Elwes measured it in 1904. When measured by Mitchell in 1989 it had grown to 24.4m × 3.9m (80ft × 12ft 10in).

Two other specimens recorded by Elwes were at Panshanger, 21.3 × 3m (70ft × 10ft) (1904): 23.7m × 3.6m (78ft × 12ft 3in) (1969) and at Badminton 15.2m × 1.5m (50ft × 5ft) (1904): 20.7m × 3m (68ft × 10ft) (1966).

The tallest Ginkgo in Britain in 1970 was a specimen at Linton Park, Kent, planted in 1844, (missed by Elwes) which according to Mitchell measured 28.3m × 2.4m (93ft × 8ft 8 in). It was blown down in 1987. All these trees are males, as indeed are most of the Ginkgos in Britain, probably because many of them have been propagated from the original Mile End tree.

Recent measurements of specimens other than those mentioned above, are:-

Carclew, Cornwall, 28m × 3.4m (92ft × 11ft 8in) 1989
Melbury, Dorset, 25m × 3.2m (82ft × 10ft 9in) 1989
Oxford Botanic Gardens, 21.9m × 2.6m (72ft × 8ft 9in) 1981
Sezincote, Glos., 29.8m × 1.6m (98ft × 5ft 7in) 1988
Whitfield Ho., Hereford, 20m × 4m (66ft × 13ft 9in) 1984

The Ginkgo is deservedly popular as an ornamental tree, not only because of its ancient lineage and the incidental immunity to disease and insect attack which this gives, but also because of its stately appearance and its attractive, strangely shaped leaves; dark green in the summer and lemon yellow in the autumn. It is not easy to get small trees established and growing vigorously because of the unpredictable way in which they put on (or do not put on) height growth; but they make pleasing subjects even in small gardens where their generally upswept branching takes up little space. Furthermore, their resistance to air pollution, smoke and dust makes them excellent trees for streets and town gardens.

It has been extensively planted as an ornamental on the Continent where it was introduced before being brought to Britain, and there are some fine old trees in most countries of Western Europe. Those in the south regularly produce viable seed.

The Ginkgo was first introduced into the USA in 1784, when William Hamilton brought a tree from England and planted it in his garden (now the Woodlands Cemetery in West Philadelphia), where it still grows and is a fine specimen. Another magnificent Ginkgo in the old Bartram Gardens could possibly be older, as this garden is itself older than that started by Hamilton.

The Ginkgo is now planted in most States, except those of the central prairies, as a street tree and in groups in parks and gardens. The fastigiate male clones are particularly popular for street planting; and Wyman mentions the one on Roosevelt Boulevard (US Route 1) in north east Philadelphia as the oldest and one of the most conspicuous.

An additional factor supporting the Ginkgo's popularity as a town ornamental, apart from its tolerance of poor soils, is its apparent immunity from insect infestation and its consequent low maintenance cost, spraying never being necessary.

As well as the fastigiate clones, pendulous varieties have been developed, and several named clones, such as 'Autumn Gold', 'Santa Cruz' and 'Lake View' are offered for sale. But, as Wyman says, 'Whether they differ one from another remains to be seen'.

Japanese Yew

Taxus cuspidata

The Japanese Yew, growing wild, is a tree sometimes as much as 18m (60ft) tall, with spreading or ascending branches. The sharply pointed needles are dark green on the upper surfaces but have the characteristic tawny-yellow bands beneath. They are arranged spirally on the branches, facing upwards. The straight trunk is clad in reddish brown bark.

Distribution and habitat

The Japanese Yew is distributed extensively in north east Asia, including parts of north east China, Manchuria, Korea and Japan. Throughout its vast range it is widely scattered, usually appearing as single specimens dotted about here and there in the forest.

In China it occurs in the mixed coniferous forests of the north eastern provinces in company with such species as Siberian Spruce *Picea obovata*, Hondo Spruce *P. jezoensis* and Manchurian Fir *Abies holophylla*; and also in mixed hardwood forests of that area, which extend northwards into Russian Sakhalin where it is found scattered amongst such trees as Siberian Elm *Ulmus pumila*, Siberian Birch *Betula mandshurica* and Amur Cork Tree *Phellodendron amurense*.

In Japan, this yew is widely distributed in Hokkaido but becomes progressively less common south through Honshu, Shikoku and Kyushu. It grows scattered through the lowland and hill forests among broadleaved and coniferous trees.

Life history

The male and female flowers occur on separate trees, appearing in early spring. The females develop into bright scarlet, shiny, berry-like fruits, borne in bunches on the branches, each fruit containing one hard, brown seed.

Growth is slow, but on favourable sites heights of 18m (60ft) and girths of 3m (10ft) are common. In northern latitudes it is often not much more than a bush.

This tree is extensively planted in Japan as an ornamental, especially in temple gardens, because of its scarlet fruits and its dense, deep green foliage. Another point in its favour as a cultivated ornamental, according to Elwes, is that even large specimens can be moved successfully.

Timber

The close-grained, reddish-brown timber is highly prized for ornamental work in Japan, and was once much used for baths, tubs, pails, water tanks, chop-sticks and clogs. Elwes found it to be milder, sounder, and freer from blemishes than *T. baccata*, and also available in much larger sizes (planks and slabs as much as 65cm (26in) wide).

Introduction, performance and use as an exotic

The Japanese Yew was introduced into Britain by Fortune in 1855, from China via Japan, but it has never grown to any size here and rarely becomes much more than a large, branchy shrub. The two specimens listed by Mitchell at Borde Hill, Sussex and Bedgebury, Kent were 5.4m × 0.5m (18ft × 1ft 11in) (1958) and 8.4m × 0.6m (28ft × 2ft) (1984). Trees of any great size are rarely seen in Britain.

The Japanese Yew was introduced into the USA in 1861 by Dr George

R. Hall who practised medicine in Shanghai and collected plants in both China and Japan. In America, the tree's performance has been much more satisfactory than in Europe. Heights of 15m (50ft) are quite common and it is one of the most widely planted coniferous evergreens in parks and gardens. It is generally hardier and more soil-tolerant than the English Yew *T. baccata*. But a hybrid between the two species, grown by T.D. Hatfield, Superintendent of the famous Hunnewell Estate in Wellesley, Massachusetts, combines the hardiness of its oriental parent with the aesthetic qualities of its English genitor to produce an excellent evergreen ornamental.

Chile Pine or Monkey Puzzle

Araucaria araucana

The Chile Pine is an evergreen, coniferous, South American tree of singular appearance. It has a circular, scarcely tapering trunk, which is clad in dark grey bark reminiscent of elephant hide. Whorls of horizontal branches emerge at regular intervals drooping at their extremities and bearing spirally-arranged, thick, sharply pointed, dark green leaves. These form a broad, columnar, domed crown and the whole tree presents a somewhat sinister, reptilian appearance. (See illustration facing p.32)

Distribution and habitat

The Monkey Puzzle has a somewhat limited natural range in two quite separate and unconnected areas; one exclusively in South Chile and the other extending just over the border into South Argentina. The former is in the coastal mountains of Araucano Province south of Concepción, and the latter some 240 kilometres (150 miles) south east in the Andes proper, the two parts extending from latitude 38°S to 40°S.

The main area of distribution, where the finest stands are found, is in the Andes. Here the trees cover vast expanses of high plateau between altitudes of 900 and 1,200m (3,000 and 4,000ft) but those growing on volcanoes in the region can be found even higher, sometimes at 3,000m (10,000ft). The forests are usually rather open and are made up of scattered stands, often on rocky ridges, with isolated Monkey Puzzles here and there. At higher altitudes there is often an understorey of Lenga *Nothofagus pumilo* and Nirre *N. antartica*, whilst lower down the Monkey Puzzles are interspersed with Raoul *N. procera* and Coigue *N. dombeyi*.

Desmond Clarke, writing in 1964 after a visit to Chile, described the unique effect these forests had upon him. 'My first sight of the Pehuen

Ginkgo, R.B.G., Kew, planted 1762 (p.28)

Chile Pine, Holker Hall, Lancs. (p.32)

Incense Cedar, Westonbirt, Glos. (p.35)

Lawson Cypress, Colesbourne, Glos. (p.37)

Nootka Cypress, Hurricane Ridge, Washington State, USA (p.39)

Hinoki Cypress, Slaley Hall, Northumberland (p.41)

Sawara Cypress var. 'Filifera', Tregothnan, Cornwall (p.43)

(the Chile Pine), as the Araucarian Indians call it, was on a cold, rainy day some 65 miles (104km) south-east of Lake Villarica, almost on the Argentine frontier. Although I knew they were to be found there, it still came as a shock to see those weird forms looming through the mist on the shore of a little lake: and seeing them again some weeks later, edging the plateau above the Laguna Malleco like sentinels in a lost world, I once again had the impression of something very remote and ancient. In the sunlight and close to, this impression is dispelled, but the araucaria is always an impressive tree. In a garden it is like an elephant in a circus.'

Life history

The male and female cones are borne on separate trees, the latter appearing, globular in shape, on the upper sides of the topmost branches and taking three years to reach their optimum size of 15 to 20cm (6 to 8in) in diameter, and ripen to shed up to 200 large seeds 2.5 to 3.7cm (1 to 1.5in) long.

Seedling growth is fairly rapid in natural stands, averaging about 45cm (18in) a year for the first 20 to 30 years. It then falls off to about 15cm (6in) a year until the century is reached when it becomes very slow, stopping altogether at about 160 years of age, although the tree may live up to 500 years. Maximum heights attained by mature trees are around 37m (120ft), with girths of up to 7.2m (24ft).

Despite its ancient appearance, the Monkey Puzzle is far from being edged out by competitors, and is especially resilient after fires, when it readily grows through *Nothofagus* scrub.

Timber

The pale brown, evenly-textured timber is used in South America for making plywood, for which it is particularly suitable, as well as for construction work and joinery; but it is in such short supply and so difficult to extract that it is not exported, unlike the Parana Pine *Araucaria angustifolia* which has a market worldwide.

Introduction, performance and use as an exotic

The Chile Pine was first seen by Europeans when a Spanish government officer, Don Francisco Dendariarena, was prospecting for ship-building timber in 1780 and came across some of these trees in Araucano Province in Chile. In 1795, Archibald Menzies, the surgeon aboard Captain Vancouver's ship *Discovery*, collected some cones (one report says that he picked up the seeds from the dinner table of the Governor of Chile) and sowed a few on board the ship. He subsequently returned to England with six plants, one of which he gave to Sir Joseph Banks and the other five to the Royal Botanic Gardens, Kew. These five were grown in a greenhouse until 1808 when one was planted out. This tree was 3.6m (12ft) tall when Loudon saw it in 1836; it died in 1892.

A further import of seed in 1844, followed by others, firmly established the Monkey Puzzle (as one of the early owners of a specimen dubbed it) as an ornamental and a curiosity in Britain, and by late Victorian times it was fashionable to plant it. Today many of these late 19th century trees are to be seen all over the country, some in the most unlikely places, towering above cottage roofs in the depth of the countryside or wedged between other trees in small urban gardens.

The last known original tree was the one which died at Kew in 1892, but a specimen at Holker, Cumbria, may possibly be another of the 1796 plantings. In 1971 Mitchell measured it as 24m × 3.3m (79ft × 11ft). It recently blew down in a gale but has been righted by cables and continues to flourish.

In an avenue of these trees, planted at Bicton, Devon, in 1844, the best of them, measured by Elwes in 1902, was 15.2m × 2.6m (50ft × 8ft 9in). In 1983 it was 28.9m × 3.6m (95ft × 12ft). Mitchell also records over 30 other specimens planted between 1830 and 1856 which range in height from 15.2 to 27.4m (50 to 90ft) and in girth between 1.8 and 3.9m (6 and 13ft), the largest being mainly near western seaboards. Some of these trees may well now have gone, but the following are specimens recently measured:-

Armadale Castle, Skye, 25m × 3.2m (82ft × 10ft 6in) 1988
Bicton, Devon, 28m × 3.9m (92ft × 13ft) (planted 1836) 1983
Dumans, Argyll, 25m × 3.3m (82ft × 11ft 1in) 1985
Powerscourt, Co. Wicklow, 25.2m × 3.3m (82ft × 11ft 2in) 1989
Wiers House, Powerscourt, Co. Wicklow, 22.5m × 3.7m (74ft × 12ft 5in) 1989
Woodstock, Co. Kilkenny, 26.8m × 3.5m (88ft × 11ft 9in) 1988

Although the Monkey Puzzle has lost much of its popularity, young specimens are still to be seen up and down the country and it seems to be better regarded in Britain than anywhere else outside its native South America. It seeds freely in western areas and in Ireland provided trees of both sexes are in reasonable proximity, and young plants are easily grown. They do not, however, grow as quickly as in their native country and may take up to 10 years to reach 90cm (3ft) in height, although growth speeds up a little later on in life.

In the USA, where it was introduced in 1795, the Monkey Puzzle has never found favour as an ornamental tree, perhaps because it grows vigorously and, according to Wyman, forms 'an ungainly mass of foliage'. It does, however, reach heights of 27m (90ft) or so and, being the hardiest of the genus, is the most planted.

Incense Cedar

Calocedrus decurrens

The Incense Cedar is a very distinctive tree in mixed conifer woodland on the western slopes of the mountain ranges of north west America. It stands on a quickly tapering, buttressed, cinnamon-red trunk which has a deeply grooved bark. The trunk supports a broadly columnar crown some 30m (100ft) in height, bearing sprays of shiny green, scale-like leaves closely pressed along the shoots. It is pyramidal in shape, but the foliage of some of the open-grown older trees reaches to the ground. (See illustration facing p.32.)

Distribution and habitat

The natural range of the Incense Cedar extends from the southern slopes of Mt Hood, in Oregon, south along the Cascade and Siskiyou Mountains to the Coast Range and the Sierra Nevada in California and even down to the Sierra de San Pedro Martir in Lower California. It is largely confined to elevations between 1,220 and 2,130m (4,000 and 7,000ft), but in the north of the range comes down to 610m (2,000 ft), and in the south may climb up to 2,440m (8,000ft). It seems to favour warm, dry hillsides but even there it is rarely found in pure stands.

In the north the Incense Cedar is found growing in association with Douglas Fir *Pseudotsuga menziesii*, Ponderosa Pine *Pinus ponderosa*, Sugar Pine *P. lambertiana*, Western Hemlock *Tsuga heterophylla*, Red Cedar *Thuya plicata*, Lawson Cypress *Chamaecyparis lawsoniana*, Tan Oak *Lithocarpus densiflora*; and in the more central parts with Jeffrey Pine *P. jeffreyi*, Big Cone Pine *P. coulteri* and Ponderosa Pine.

Life history

The Incense Cedar produces masses of little urn-shaped cones (immediately noticeable as being different from round ones of the otherwise superficially similar cypresses) hanging on the ends of the branchlets. Seed starts to fall in late summer. The seedlings develop best in partial shade but they are slower growers than most of their associates, taking three to five years to reach a height of 15cm (6in). The growth rate then increases reaching a maximum between 50 and 75 years, but never catching up with competitors.

In the Sierra Nevada the maximum height is usually about 45m (150ft) with girths of 6m (20ft). In Oregon a tree 68m (225ft) tall was reported – an exception for this species. The tree is certainly more than 500 years old; the oldest recorded being 542 years for a tree which girthed only 4m (13ft). Old trees have an extensively developed lateral root system and are thus very wind-firm.

The Incense Cedar is more shade-tolerant than Douglas Fir and the associated pines, but less so than Grand Fir. Seedlings can endure intense shade and respond to opening up with vigour, but with their growth rate being so slow, they are destined always to be a secondary species in any stand.

Timber

The soft, light brown timber is straight-grained, cleaves easily and does not check or warp during seasoning. It is used for shingles, sleepers, doors and window frames; and because of its aromatic odour it is used for making cedarwood chests.

Introduction, performance and use as an exotic

The Incense Cedar was discovered in 1846, near the upper waters of the Sacramento River in California. It was introduced into Britain by Jeffrey in 1853. A number of original trees are still alive, one of these, which Elwes noted as the finest in England and of which he provided an excellent photograph, is at Frogmore. It was planted in March, 1857 and was 19.8m × 2.7m (65ft × 9ft) when Elwes measured it in 1904.It had grown to 30.1m × 4.6m (99ft × 15ft 6in) at 15cm (6in) when Mitchell measured it 60 years later. Other trees mentioned by Elwes and later measured by Mitchell are of interest. One is at Keir House, Perthshire which was planted in 1859. By 1904 it measured 12.8m × 1.4m (42ft × 4ft 10in); 66 years later it had grown to 21.3m × 1.9m (70ft × 6ft 10in). Another at Killerton, Devon, of unknown planting date, measured 32m × 3.5m (105ft × 11ft 3in) in 1983.

The tallest specimen in Britain in 1971, according to Mitchell, was at Oakly Park, Hereford. It measured 36m × 3.2m (118ft × 10ft 6in) Other trees have now overtaken it as recent measurements show:-

Belladrum, Inverness, 26.8m × 4.5m (88ft × 15ft 2in) 1988
Charleville, Co. Wicklow, 29.8m × 4m (98ft × 13ft 11in) 1989
Doune House, Perth, 37.8m × 5.8m (124ft × 19ft 4in) 1988
Endsleigh, Devon, 39.6m × 4m (130ft × 13ft 8in) 1987
Forde Abbey, Dorset, 32m × 3.5m (105ft × 11ft 9in) 1988
Mt Usher, Co. Wicklow, 35m × 3.8m (115ft × 12ft 6in) 1989

The exceptionally narrow, columnar habit of these trees in Britain, apart from being one of their attractive features, is perplexing, especially as it seems to be more accentuated the further east the tree is planted. In their natural habitat the adults are broadly columnar and the young trees distinctly pyramidal. When planted in Ireland they become less broadly columnar, and in the east of England extraordinarily narrow. Young trees in Britain are never pyramidal, always narrow columns. Because of this trait and because open-grown trees hold their shiny green, feathery foliage down to the ground, they are very popular subjects for garden planting. Small groups of them, narrow and formal, and surrounded by maples, as at Westonbirt, Glos., form a remarkably arresting picture.

Although the Incense Cedar ripens seed on the Continent it rarely does so in Britain. Italy has some of the largest specimens found in Europe.

In the USA this native tree is very popular for formal planting, both because of its shape and its immunity to insect attack. On good moist soil it quickly becomes a splendid ornamental tree.

Lawson Cypress

Chamaecyparis lawsoniana

The Lawson Cypress is a very remarkable tree, known worldwide but with a very limited natural range in the north west Pacific coastal region of America, where it is found in groups or scattered about in the great conifer forests. The buttressed trunk, clad in thick, reddish-brown, fibrous bark, reaches up in the dense forest 60m (200ft) or more, often free of branches for 45m (150ft), to a crown of delicate, flattened, feathery leaves with a narrow, spire-like head. In open forest these cypresses, narrowly conic in shape, are furnished to the ground with their dark green fronds, the lowest branches sometimes arching down and touching the ground. (See illustration facing p.32.)

Distribution and habitat

The natural range of the Lawson Cypress is restricted to 210 kilometres (130 miles) of the Pacific coastline from Coos Bay, Oregon, in the north to Mad River, north California, in the south. It is rarely found more than 64 kilometres (40 miles) inland, except for a few isolated pockets in the Siskiyou Mountains and one or two scattered locations on Mount Shasta and the Trinity Mountains in north California.

This cypress is most abundant round Port Orford (the usual American name is Port Orford Cedar) and it prefers valley bottoms and moist hillsides. But it also grows as high up as 1,530m (5,000ft) on dry sandy ridges in the mountains, and on soils which are generally unfavourable to most tree growth.

In the coastal areas Lawson Cypress associates with Douglas Fir *Pseudotsuga menziesii*, Western Hemlock *Tsuga heterophylla*, Grand Fir *Abies grandis*, Sitka Spruce *Picea sitchensis*, Lodgepole Pine *Pinus contorta*, Western Red Cedar *Thuya plicata* and Redwood *Sequoia sempervirens*. Inland on the higher ground it is found with Western White Pine *Pinus monticola*, Sugar Pine *P.lambertiana*, Incense Cedar *Calocedrus decurrens*, Ponderosa Pine *P. ponderosa*, Jeffrey Pine *P. jeffreyi*; and in the Siskiyous with Brewers Spruce *Picea brewerana* and Sadler's Oak *Quercus sadleriana*.

Life history

Although some trees start to bear cones at eight years of age, 10 is more usual, reaching maximum production at 100 years. Most seed germinates as soon as it falls, but some may remain dormant for up to five years. Seed production is enormous, a fall of up to 1.75 million seeds to the acre (0.4 hectares) has been recorded. The seedlings grow slowly at first – 7cm (3in) in two years – but later growth becomes more rapid, averaging about 25cm (10in) a year, provided there is not excessive shade, although this cypress is well able to live beneath the canopies of most associates. In such mixed stands, Douglas Fir and other overstorey trees tend to die out gradually after 250 to 300 years, releasing the more persistent Lawson Cypresses which have a great capacity to respond to the opening up of the canopy. After 400 years or so the stand may be composed largely of the cypresses, with an understorey of shade-bearing hemlocks and younger Lawsons.

Top heights of mature Lawsons are often in excess of 60m (200ft), with girths of 3.5 to 4m (12 to 14ft). Such trees will usually be 300 to 350 years old but some of the largest may be as much as 600.

Timber

The timber of Lawson Cypress is even-grained, white, moderately soft and durable, easily worked and capable of taking a good polish; all of which make it very desirable were it not that the tree's very limited range makes supplies scarce. It is used for boats, plywood and flooring; and its ginger-like odour makes it valuable for lining drawers, wardrobes etc.

Introduction, performance and use as an exotic

Lawson Cypress was introduced into Britain by William Murray who sent seeds from the Upper Sacramento Valley to Messrs Lawson's nursery in Edinburgh in 1854. From these, only four plants were raised, but one of these survived until 1955 at least, when Mitchell measured it to be 26m (85ft) tall and 3m (10ft) in girth. A previous measurement, in 1931, gave it as being 17m × 1.8m (56ft × 6ft 6in). In 1855 a much larger consignment of seed was sent by Murray and from then on the tree never looked back. This cypress was planted all over the country, in estates, and in gardens large and small. Some of these early trees have now reached heights of 30m (100ft) and more in as many years, but most of them have disappeared, probably due to windblow, for those which remain are all very healthy. Recent measurements of some of the earlier trees include:-

Arkinglass, Argyll, 41m × 2.5m (135ft × 8ft 5in) 1989
Balmacaan, Inverness, 40.2m × 4.2m (132ft × 14ft) 1987
Doune House, Perth, 37m × 4m (121ft × 13ft 9in) 1988
Powerscourt, Co. Wicklow, 31m × 4.9m (102ft × 16ft 4 in) 1989
Stonefield, Argyll, 26.8m × 3.7m (88ft × 12ft 3in) 1986

(the tree at Endsleigh, Devon, which in 1980 measured 40.5m × 3m (133ft × 10ft 3in), has lost its top)

The most remarkable thing about this cypress is its tendency to vary in cultivation in Europe but not in its native land. Seventy or more cultivars have been named of which about half grow into trees of between 6 and 21m (20 and 70ft) tall, and varying in colour from blue to gold and yellow; and in shape from tall, erect pillars to feathery, cascading columns. Others are dwarf and creeping plants which hug the ground. Equally remarkable is the universal hardiness of the type and the varieties of a tree which hails from the mild climate of the Pacific coast.

This cypress is not often planted as a forest tree. Its height growth is not rapid compared with other north west American conifers, and its tendency to fork at an early age spoils its form. Even so, its shade-bearing qualities and its narrow habit do give opportunities for greater stocking and thus a possible place on certain sites.

In New Zealand, this cypress was at one time extensively planted in shelterbelts, especially in North Island, but disease has resulted in its replacement by other species, in particular Leyland Cypress *Cupressocyparis leylandii.*

In the USA, where very few variations have occurred and where only about 30 varieties are offered for sale (mostly originating in Europe), Lawson Cypress, in its more colourful and elegant cultivars, is a popular garden and park tree, especially in the milder, moister north western states.

Nootka Cypress

Chamaecyparis nootkatensis

The Nootka Cypress or Alaska Cedar is one of the smaller conifers of the cool, humid areas of north west America, where winter snowfall is heavy. Its conical crown of flexible drooping, feathery branches quickly throws off the gathering snow. At best this cypress is rarely more than 30m (100ft) tall. The broad, buttressed and often fluted bole tapers rapidly and is covered in ashy-brown fibrous bark which flakes off in strips. It is often clean of branches for half its length. (See illustration facing p.33.)

Distribution and habitat

The natural range of the Nootka Cypress stretches from Prince William Sound in northern Alaska (61°N) down through British Columbia, Washington and Oregon (42°N). In the northern part, it is mainly confined to coastal areas, but in southern British Columbia, on Vancouver Island and on the western slopes of the Cascades, it is found at altitudes between 760 and 1,830m (2,500 and 6,000ft).

Provided there is an abundance of moisture, the Nootka Cypress grows on the meanest of soils – boggy, rocky and peaty – and although it fares well on the deeper soils of the valley bottom, it is rarer on these sites because it cannot compete with other conifers, such as Douglas Fir and Western Hemlock, which also grow there.

This cypress is found growing in association with Mountain Hemlock *Tsuga mertensiana*, Sub-alpine Fir *Abies lasiocarpa*, Sitka Spruce *Picea sitchensis*, Western Hemlock *Tsuga heterophylla* and Western Red Cedar *Thuya plicata*; also sometimes in pure stands where the spruce and hemlock have been destroyed by fire.

Life history

The very small, spherical cones take two years to ripen, with heavy crops every two to four years, and the seedlings are very slow growers. Indeed, at all stages of its life this cypress grows very slowly, more so than most other western conifers, although it long outlives most of them. One stump which was hollow and 5.4m (18ft) in circumference showed 1,040 annual rings in the 30cm (one foot) thick outer shell. Some trees may live for 3,500 years.

The largest known Nootka Cypress is in the Olympia National Park, Washington; it is 53m (175ft) tall and 6.4m (21ft) in girth.

This cypress is moderately shade-tolerant; more so than Western White Pine and Noble Fir *Abies procera* but less so than Western Red Cedar and Western Hemlock. This tolerance tends to lessen as it ranges further north where younger stands are essentially of the same age. With its slow growth rate it can compete with hemlock and cedar only on sites less favourable to these two species.

Timber

The very limited supply of this heavy, strong, durable, yellowish timber tends to limit it to those specialist uses to which it is ideally suited, such as boat-building, canoe paddles, patterns and cabinet work.

Introduction, performance and use as an exotic

The Nootka Cypress was discovered by Archibald Menzies in 1793, but it does not appear to have been cultivated in Europe until 1850 when some seedlings were raised in St Petersberg Botanic Garden. It came to Britain in 1854 when Pontey of Plymouth received some 90cm (3ft) tall plants from Booth of Hamburg.

A tree at Eastnor Castle, Herefordshire, which may be one of the first batch, was reported by Elwes as being 17.3m × 1.3m (57ft × 4ft 6in) in 1909. Mitchell measured this same tree in 1988 and found it to be 26.8m × 3.2m (88ft × 10ft 8in). The tallest found by Elwes in the first decade of this century was a specimen at Eggesford, Devon, which measured

18.6m × 1.6m (61ft × 5ft 7in), and this again was probably one of the original trees. Mitchell's measurements of some 42 specimens up and down the country show the tallest, in 1970, to have been at Doune House, Perthshire; it was 29.8m × 2.4m (98ft × 8ft). Others have outgrown it, as some later measurements show:-

Blair Drummond, Perths., 31m × 3m (102ft × 10ft 2in) 1986
Burns House, Kincards., 25m × 3.2m (82ft × 10ft 6in) 1987
Fairbairn, Ross & Cromarty, 26.8m × 2.9m (88ft × 9ft 7in) 1986
Penrhyn Castle, Gwynedd, 22.8m × 3m (75ft × 10ft 1in) 1989
Stourhead, Wilts., 28m × 2.9m (92ft × 9ft 8in) 1987
Westonbirt, Glos., 31m × 2.8m (102ft × 9ft 2in) 1989.

The Nootka Cypress does not seem to fare well in continental Europe but is hardy in Norway as far north as Trondheim.

As an ornamental in Britain it is exceptionally elegant and hardy, growing well on both acidic and calcareous soils, and it is frequently to be found in parks, gardens and churchyards, although much less common than Lawson Cypress.

As a forest tree it has been little planted, largely because of its very slow growth in early years. Such plantations as have been formed, at best grew just 1.8m (6ft) or so in 20 years and at worst a mere 90cm (3ft) in 16 years.

The Nootka Cypress is one of the parents, with the Monterey Cypress *Cupressus macrocarpa*, of the Leyland Cypress *Cupressocyparis leylandii*.

In the USA this native cypress does not seem to be widely planted as an ornamental except in the north west, where it is indigenous and where the climate is moist. There, its long, drooping sprays make it a popular subject although its ultimate size (up to 37m (120ft) tall) renders it unsuitable for town planting.

Hinoki Cypress

Chamaecyparis obtusa

The Hinoki Cypress is one of two cypresses exclusively indigenous to Japan. From an unusually straight bole, clad in notably reddish bark, which peels off in strips, the ascending branches carry the usual scale-like leaves of a cypress. However, these are smaller and much blunter than other species, and are of a particularly bright, shiny green. (See illustration facing p.33.)

Distribution and habitat

The natural range of the Hinoki Cypress is confined to the central and southern parts of Japan – south Honshu and Shikoku, and scattered areas in Kyushu – at elevations of between 610 and 1,530m (2,000 and 5,000ft). Although often found growing in more or less pure stands on ridges and on upper slopes, in more sheltered sites at higher altitudes this cypress is associated with Alcock's Spruce *Picea bicolor* and with Hondo Spruce *P. jezoensis* var. *hondoensis*. Lower down the mountains it grows alongside trees such as the Keaki *Zelkova serrata* and various species of oaks, maples and chestnuts.

Even though it is limited to central and southern parts in its natural state, the Hinoki Cypress has, since ancient times, been planted both for its valuable timber and as one of the 'Five Trees of Kiso', protected under Japanese law as the valuable species in the Forest of Kiso. The other four were the Sawara Cypress *Chamaecyparis pisifera*, the Umbrella Pine *Sciadopytis verticillata*, the Hiba *Thuyopsis dolobrata* and the Japanese Red Cedar *Thuya standishii*.

Life history

The little, round female cones, barely 12mm (0.5in) in diameter, are borne at the ends of the branches, often in such profusion as to form clusters. They release their seeds in the autumn of the first year.

Growth is slow and in maturity the Hinoki Cypress does not usually attain a height of more than 27m (90ft) with a girth of 3m (10ft) or so. But there have been reports of trees 42m (140ft) tall drawn up by other species surrounding them. This usually happens on good soils in more sheltered spots at lower altitudes.

Timber

The timber of this cypress is one of the most valuable softwoods grown in Japan. White to straw-coloured, it is light, strong and tough, the fine grain giving it a silky lustre. It is greatly valued for building temples and houses, as well as for interior work and as a basis for the fine lacquer work for which the Japanese are renowned.

Introduction, performance and use as an exotic

The Hinoki Cypress was introduced into Britain by J. Gould Veitch, the Exeter nurseryman, in 1861.

The largest tree which Elwes knew of was at Bicton, Devon. It measured 11m × 1.2m (36ft × 4ft 4in) in 1906; and Mitchell found it to be 24m × 2.4m (79ft × 8ft 2in) in 1968. In 1966 he recorded the tallest specimen as being one at Bedgebury House, Kent, which measured 25.3m × 2.5m (83ft × 8ft 5in). In 1977 he found it to have grown to 26.5m × 2.6m (87ft × 8ft 8in). Other trees more recently measured are:-

Culdeas, Perth, 18.9m × 2.4m (62ft × 8ft) 1986
Endsleigh, Devon, 24.4m × 2.1m (80ft × 7ft 4in) 1987
Killerton, Devon, 24m × 1.9m (79ft × 6ft 6in) 1983

Strone House, Perth, 21.9m × 2.5m (72ft × 8ft 3in) 1989
Tregrehan, Cornwall, 15.8m × 2m (52ft × 6ft 9in) 1987

It is for the extraordinary number of varieties and cultivars of different shapes, sizes and colours that this cypress is so widely planted in gardens in Europe and the United States. All seem to be perfectly hardy and tolerant of most soils. The golden cultivar 'Crippsii', originating in Tunbridge Wells, Kent in 1901, is one of the most popular and is a dense, broadly conic tree; some of the oldest specimens have attained heights of over 15m (50ft). Others include 'Filicoides', which is a slender tree with dense feathery foliage growing to 12m (40ft) tall, originating in Japan and brought to Britain at the same time as the type tree. Another is 'Lycopoides' introduced again at the same time from Japan. This has blue-white markings on the bright green leaves and grows slightly taller than 'Filicoides'. There are also many dwarf forms, attractive for rock gardens, and the type tree is a favourite for dwarfing and bonzai work.

In the USA where the tree was introduced in 1861, the type is not often seen in the eastern states, but everywhere the shrubby forms are much more popular, especially the fastigiate form 'Erecta' as well as 'Filicoides' and 'Gracilis'.

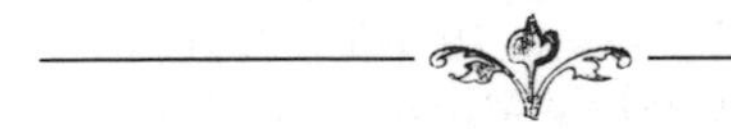

Sawara Cypress

Chamaecyparis pisifera

The Sawara Cypress is the other of two exclusively indigenous cypresses of Japan. In almost all respects, bark, foliage colour and size, it is very similar to the Hinoki Cypress, but in one important particular it differs. The scale-like leaves have fine, incurved tips and each is white at the base. In general, the Sawara is more pendulous and graceful than the Hinoki. (See illustration facing p.33.)

Distribution and habitat

Like the Hinoki, the natural range of the Sawara is confined to the central and southern parts of Japan; fairly widespread in southern Honshu and in all of Shikoku, but scattered in Kyushu. Whereas the Hinoki is mainly confined to the upper slopes and ridges growing as high up as 1,830m (6,000ft), the Sawara tends to favour the lower, damper sites near mountain streams. Here, if Hinoki is present at all, it is definitely the less numerous and grows in company with deciduous species such as oaks, maples and chestnuts.

Life history

The female cones are even smaller than those of the Hinoki, barely 6mm (0.25in) in diameter and green, bloomed with white, early on, but ripening to a deep brown in the autumn to shed their seeds.

Growth appears to be rather slow, not much more than 30cm (1ft) a year. The largest tree which Elwes saw in Japan measured 35m × 3.9m (115ft × 13ft); it was about 150 years old. In the dense forest, the Sawara Cypress can have a trunk up to 24.4m (80ft) to the first branch and a total height of 36.6m (120ft) or more. The largest on record in Japan was 40.8m × 6.4m (134ft × 21ft). A stump which Elwes saw had 200 countable rings in a diameter of 1.2m (4ft), but the tree had probably been over 250 years old when felled.

Timber

The timber, which is very fine-grained, is not so much valued as that of the Hinoki but is used for many of the same purposes.

Introduction, performance and use as an exotic

The Sawara Cypress was first introduced into Britain in 1861 by J. Gould Veitch. The finest specimen which Elwes saw, measured and photographed was at Bicton, Devon; it was 12.5m × 1.4m (41ft × 4ft 10in) in 1906. A tree at Bicton measured by Mitchell in 1968 was 23.7m × 1.5m (78ft × 5ft), and the tallest in Britain; it might have been the same tree that Elwes listed. By 1977 it had grown to 25m × 1.7m (82ft × 5ft 9in). The trees which Elwes mentions as being at Dropmore, Bucks., and at Killerton, Devon, could be those which Mitchell records as being 21.3m × 3.4m (70ft × 11ft 7in) and 22.8m × 2.5m (75ft × 8ft 4in) respectively in 1987. Some recent measurements include the following:-

Cowdray Park, Sussex, 28m × 2.7m (92ft × 9ft 1in) 1984
Endsleigh, Devon, 21.9m × 2.2m (72ft × 7ft 3in) 1987
Killerton, Devon, 22.8m × 2.5m (75ft × 8ft 4in) 1987
Strone House, Perth, 22.8m × 2.3m (75ft × 7ft 9in) 1989
Tottenham House, Wilts, 22.8m × 2.4m (75ft × 8ft) 1984

If the type tree is comparatively rare, the cultivars, which are very numerous, are extremely common in Britain, making this species the most favoured cypress in cultivation after the Lawson. One of the most frequently planted cultivars is 'Plumosa' which came from Japan in 1861. It retains its feathery juvenile foliage throughout its life to make a dense domed crown. There is an original tree of this cultivar at Scorrier House, Cornwall, planted in 1867, which measured 16.7m × 2.5m (55ft × 8ft 8in) in 1959; and there was another at Linton Park, Kent, planted in 1866, which grew to 20m × 2.8m (66ft × 9ft 3in) but died in 1956.

Another commonly planted cultivar is 'Squarrosa', which originated in Japan and was introduced into Britain as early as 1843, before the type tree. It has blue-grey, fluffy shoots of juvenile foliage and there is a fine specimen at Linton Park, planted in 1866 which, in 1956, measured 20.7m × 1.5m (68ft × 5ft). Most of the others listed by Mitchell are between 15.2 and 19.8m (50 and 65ft). The cultivar 'Filifera aurea' forms a low cone rarely more than 9m (30ft) tall. The shoots of adult foliage develop into long, hanging threads of a golden colour.

These three cultivars are also extensively planted across the United States, as well as another, less frequently seen in Britain, called 'Aurea', a tree like the type in shape and size but with yellow new shoots which make it very attractive in summer. Mulligan says that the type makes an excellent town tree in the north-western States.

There are many other cultivars, most of them dwarf or semi-dwarf, which are very popular for planting in rock gardens in Europe and America. Outside Japan this cypress has failed as a forest tree.

Monterey Cypress

Cupressus macrocarpa

In its extremely restricted natural range, the smallest of any conifer in North America, the Monterey Cypress is, for the most part, a gnarled, twisted, flat-topped tree, buffeted by the wind and clinging to rocky seaside cliffs. Inland from the immediate Pacific Coast, where the trees are young and crowded, the crowns are rigidly straight, but eventually flatten out, even the tallest trees rarely exceeding 18m (60ft) in height. The bark of the trunks and the enormous, spreading lower branches, is deep red-brown, weathered here and there to a deep grey; and the stout twigs in the crown are covered with scale-like overlapping leaves. (See illustration facing p.48.)

Distribution and habitat

The natural range of the Monterey Cypress is a narrow Pacific coastal strip in California, just 3 kilometres (two miles) long, from Cypress Point to the south side of Carmel Bay, and also on Point Lobos.

Why a tree which grows with such vigour when planted almost everywhere else in the temperate world should grow so meanly in such a limited area, is difficult to understand. Possibly, prior to the last glaciation, it was more widespread inland, but became isolated on the Monterey Peninsula which was free of ice, and never migrated back again when the ice retreated. Or it may have been more widespread in the area which is now beneath the ocean.

Its sole associate is the Monterey Pine *Pinus radiata* with which it mingles on the inland edge of its narrow domain.

Life history

The large round to oblong, ashy-brown cones 3.8cm (1.5in) long, are clustered in bundles on the branches, shedding their seed in the late summer of their second season.

The seedlings which find favourable sites grow rapidly and maintain their vigour until their height of 15 or 18m (50 or 60ft) is reached, when growth slows down and the crowns start to flatten out. Trees with boles girthing 90 or 120cm (3 or 4ft) are often 60 to 70 years old, and some of the bigger specimens in the natural habitat are thought to be at least 200.

Timber

The yellow-brown timber, with rose-red streaks and a cedar-like odour, is fine-grained, heavy, hard and very durable; but it is of no commercial value in America.

Introduction, performance and use as an exotic

The Monterey Cypress was first cultivated in the Horticultural Society's garden at Chiswick in 1838, grown from seed of unknown origin presented by A.B. Lambert. Two or three years later some plants were raised in Low's Nursery at Clapton from seeds sent from St Petersburg Botanic Gardens taken from what was described as 'a new species of cypress from California'. Hartweg actually discovered the tree near Monterey in 1846.

Of these earliest trees eight, and probably more, are still living. Elwes gives a photograph of a specimen, planted by the Prince Consort in 1846, which he found to be 23.7m × 4.8m (78ft × 16ft) in 1906. In 1964 Mitchell measured it as being 25.9m × 6.9m (85ft × 23ft). Yet, one planted at Hamwood House, Co. Meath, in 1844, which Elwes recorded as 25.9m × 3.2m (85ft × 10ft 6in) in 1903, was still the same height but over 6.3m (21ft) in girth when Mitchell measured it in 1968. Elwes also illustrates a very well furnished specimen at Beauport which was 19.5m × 5m (64ft × 17ft) at 90cm (3ft) in 1905. It had grown to 25.9m × 8m (85ft × 27ft) at 30cm (1ft) when measured by Mitchell in 1965. There are two trees at Bicton, Devon, reputedly planted pre-1846, which according to James were 34.4m × 3.9m (112ft × 13ft 2in) and 33.8m × 4m (111ft × 13ft 6in) when he measured them in 1969.

The tallest specimen in Britain in 1965, according to Mitchell, was at Tregothnan, Cornwall which topped 36.6m (120ft) and girthed nearly 6m (20ft). But more recent measurements show others taller than this:-

Birr Castle, Co. Offlay, 31m × 5.8m (102ft × 19ft 3in) 1989
Johnston Castle, Co. Wexford, 40m × 8m (132ft × 26ft 4in) 1989
Melbury, Dorset, 40.4m × 6.1m (130ft × 20ft 5in) 1989
Montacute House, Somerset, 36.6m × 7.6m (120ft × 25ft 5in) (planted in 1840 and one of the originals) 1987
Moreton House, Bideford, 40m × 9.4m (132ft × 31ft 3in) 1986

Torlosk, Mull, 35m × 6.1m (115ft × 20ft 6in) 1987
Westonbirt, Glos., 36m × 3.9m (118ft × 13ft) 1989

Despite its natural home being a place where frosts never occur, the Monterey Cypress is very hardy and in Britain is only damaged by severe frost or killed by the severest of winters. It is very commonly planted in most lowland areas of Britain where its evergreen, feathery foliage, its neat habit, its quick growth and its tolerance of most soils make it a popular garden tree. Furthermore, it takes kindly to clipping and makes an excellent hedge, especially in seaside areas where it withstands the worst of salt-laden gales.

A number of cultivars such as the spire-like 'Fastigiata', and 'Lutea', 'Donald Gold' and 'Goldcrest' which are varying shades of yellow, are now widely planted.

Because of its susceptibility to damage in upland areas where exposure is severe – the very areas where exotic conifers are most extensively planted in Britain – this cypress is rarely planted as a forest tree.

In the USA this native cypress is extensively planted in the milder coastal areas as a hedging plant.

One of the most noticeable features of the New Zealand countryside (and in parks in the cities) is the preponderance of this cypress, often planted a century ago as shelter for farmsteads and stock. Many of the trees are of enormous size, usually multi-stemmed and up to 30m (100ft) tall. Their size renders them prone to windfall and this is likely to be the eventual fate of most. But their rapid growth rate to great sizes and their resistance to salt-laden winds makes them unrivalled shelter trees in New Zealand where strong winds are a frequent occurrence.

Smooth Arizona Cypress

Cupressus glabra

The Arizona Cypresses come from high up in the Arizona mountains where they stand on canyon sides and mountain slopes, often in groups of considerable size, with upswept branches like narrow cones. They are 15 to 18m (50 to 60ft) tall, blue-grey in colour and feathery of form; in both respects quite distinct from associated trees. The quickly tapering boles have a reddish-brown bark with flaking circular scales.

Distribution and habitat

The natural range of the Arizona Cypress is the mountains of central southern and eastern Arizona where it grows on gravelly slopes and in moist places, as well as on exposed dry sites on canyonsides and rocky mountain slopes at altitudes of between 1,220 and 2,140m (4,000 and

7,000ft). It is very often in pure stands but sometimes in mixture with Limber Pine *Pinus flexilis* and live oaks.

Life history

The male cones are bright yellow and very numerous, often making the trees look tinged with gold in winter. The round female cones, with prickly projections, scatter their seed in September of the second season. The seed crops are large, but so many of them fall on stony ground that survivals are few due to drought or sudden rainfall carrying away the seed. Growth is slow, especially at higher altitudes, and trees of 90 to 120cm (3 to 4ft) in girth may be anything from 100 to 400 years old.

Timber

The timber is mostly inaccessible and in short supply. Where it is available it is used for grape stakes and bean poles.

Introduction, performance and use as an exotic

There was originally some confusion over *Cupressus arizonica* and *Cupressus glabra*, the latter being introduced into Britain as the former in 1882 from the Arnold Arboretum in Washington, D.C. But the distinction was noted and correctly named imports were started in 1907.

A few of the trees mentioned by Elwes as *C. arizonica* were probably *C. glabra* and do not now seem to be alive.

In 1960, Mitchell recorded a number of large specimens of which the tallest was at Sheffield Park, Sussex, measuring 21m (70ft) in height. He also records the growth rate of some trees which exceeds or comes near to 30cm (1ft) a year for the first 30 to 40 years.

Recent measurements show some specimens now exceed the 1960 record height by many metres:-

Bulkeley Mill, Gwynedd, 18.9m × 2m (62ft × 7ft 3 in) 1984
Henrietta Park., Bath, 21.3m × 1.2m (70ft × 4ft 4in) 1984
Mt. Usher, Co. Wicklow, 16.7m × 1.8m (55ft × 6ft) (*arizonica*) 1989
Powerscourt, Co. Wicklow, 24.4m × 2.4m (80ft × 8ft) (*arizonica*) 1989
Victoria Park, Bath, 24.4m × 1.7m (80ft × 5ft 8in) 1989
Wisley, Surrey, 22.8m × 1.8m (75ft × 6ft 1in) 1987

It is interesting to note that the Powerscourt tree, measured by Mitchell as 16.1m × 1.8m (53ft × 6ft 2in) in 1966, has apparently increased in height by 8m (27ft) in 23 years.

The commonest and most handsome of the Arizona cypresses now grown in many gardens up and down the country, where they are universally hardy, is the cultivar 'Pyramidalis' which has intensely blue-white foliage, giving its attractive, upswept, conic crown an exceptional beauty.

Monterey Cypress, Point Lobos, Monterey, Calif. (p.45)

Italian Cypress, Kolossi, Cyprus (p.49)

Himalayan Cypress, Scorrier, Cornwall (p. 51)

Chinese Juniper, Tai Lam Chung, Hong Kong (p.53)

Chinese Thuya, Punjab, India (p.55)

White Cedar, Weston-under-Lizard, Salop. (p.57)

In the USA this native cypress, sold by the trade as *Cupressus arizonica* 'Bonita' and similar if not identical with the cultivar 'Pyramidalis', is planted in gardens as a handsome and hardy ornamental. According to Wyman, however, it often has a poor root system and is easily blown over. Nurserymen have thus taken to grafting it onto other rootstock.

Italian Cypress

Cupressus sempervirens

The Italian Cypress is an evergreen tree of varying habit and often of very considerable size. The variety 'Horizontalis' has a broad, spreading form with a flat top to the crown in old age, something like a cedar; the variety 'Stricta' forms a thin, slender column, both growing together in the wild, but the former much the more numerous. The bole has a thin greyish-brown bark with longitudinal fissures, and the small, scale-like, dark green leaves are closely pressed on the twigs. (See illustration facing p.48.)

Distribution and habitat

The natural range of the Italian Cypress extends from Greece, Crete and Cyrenaica eastwards through the islands of Rhodes and Cyprus to the Taurus Mountains and the mountains of northern Persia and south into the Levant including Syria, Lebanon, Israel and Jordan. It is found at elevations varying from sea level to 1,220m (4,000ft), favouring limestone soils. Although pure stands do occur, mostly at higher elevations, this cypress is more usually found in mixture with Aleppo Pine *Pinus halepensis* and Stone Pine *P. pinea*, as well as with Maritime Pine *P. pinaster* at lower levels in the west of the range.

This cypress, especially the upright variety, has become naturalised in many places in southern Europe, particularly in Italy where it grows in mixed woods with oak and ash. It is planted extensively throughout the Mediterranean region, where its tall, green columns contrast very pleasingly with the white, flat-topped buildings.

Life history

The round, shiny green cones, 2.5cm (1in) in diameter, are borne in clusters, turning brown as they start to ripen in the autumn of the first year, but remaining unopened on the tree to shed their seeds in the following autumn. Growth is rapid in the early stages, with shoots up to 60cm (2ft) long; but in later years it becomes very slow and finally almost imperceptible. This tree is apparently very long-lived. Elwes writes of a specimen at Summa in Lombardy about which chronicled evidence existed purporting to prove that it was planted before 42 BC. Manetti measured it for Loudon in 1838 and found it to be 25.6m × 5m (84ft × 17ft). Elwes saw it in 1906, when it was covered with cones and measured 37m × 8m (121ft × 27ft), but he doubted the story about its reputed age.

The cypresses at the Diocletian Museum in Rome are said to have been planted by Michaelangelo. When measured in 1817 at presumably over 300 years of age, the largest girthed 5m (17ft). They are of the upright form. Another very large specimen was one which grew in Greece, near Mistra. It was over 45m (150ft) tall and girthed 11m (36ft) in 1839, but was destroyed by fire in 1881.

Although reports of great ages may be exaggerated, there is no doubt that Mediterranean Cypresses are very long-lived, and the flat-topped variety attains enormous girths.

Timber

The moderately hard, close-grained, fragrant and very durable timber has been prized since ancient times. The gopher wood, of which the Ark of the Covenant was built, was thought to be from this cypress; its durability is legendary. Pliny tells of the statue of Jupiter in the Capitol which was made of cypress timber and, even after 600 years, showed no signs of decay. The doors of the Temple of Diana at Ephesus, made of this wood, still looked like new when over 400 years old. The doors of St Peter's, Rome, also supposedly of cypress timber, were 1,000 years old and still quite sound when they were removed. Nowadays the timber is used for construction work, house building, furniture and for clothes chests.

Introduction, performance and use as an exotic

The exact date of introduction of the Italian Cypress is unknown but Turner, who was a physician at Syon in 1548, says that there were many of these trees there at that time. Loudon mentioned one of them as being the largest and the oldest in his time, and probably one of those which Turner planted during the reign of Henry VIII. He mentioned another, then 'in ruins', at Ditton near Windsor, which was planted by Cardinal Wolsey. But none of the Syon trees nor the one at Ditton is mentioned by Elwes and presumably, like many other large specimens, they had succumbed to the harsh winters of 1860-61 or 1879-81. The finest specimen known to Elwes was one which he measured and photographed at Heron's Court, Richmond, in 1906. It was 18.3m × 1.5m (60ft × 5ft) and of the upright type. It had gone by 1950 when Gardner looked for it. But a tree at Killerton, Devon, planted in 1841, which Elwes measured as 15.2 to 18.3m (50 to 60ft) in 1906, was still there in 1970 when Mitchell found it to be 18.9m × 1.9m (62ft × 6ft 5in). Probably the oldest Italian Cypress now living in Britain is at Dropmore, Bucks. It was planted in 1800 and was 12.8m × 1.6m (42ft × 5ft 9in) when Elwes measured it in 1906, but only 14.6m × 1.8m (48ft × 6ft 1in) when Mitchell saw it in 1957. In 1970 he recorded the tallest and largest of this cypress, both at Nettlecombe Court, Somerset, as being 21.3m × 1.8m (70ft × 6ft 7in) and 20.4m × 3.7m (67ft × 12ft 7in) respectively. Some more recent measurements are:-

Biel, E. Lothian, 17m × 2.8m (56ft × 9ft 5in) at 60cm (2ft) 1985
Blenheim Palace, Oxon., 20.7m × 2.4m (68ft × 7ft 11in) 1980
Borde Hill, Sussex, (Warren) 22.5m × 1.1m (74ft × 2ft 10in) 1987
Killerton, Devon, 21.3m × 1.8m (70ft × 6ft 4in) 1980
Powerscourt, Co. Wicklow, 23.4m (77ft) 1989
Singleton Abbey, Glam., 20m × 1.8m (66ft × 6ft 4in) 1989

Although Loudon was able to write, in 1838, that 'there is scarcely a suburban villa or a country seat in which it is not to be found', this cypress is not now common in Britain. It is found mainly in the largest gardens in the south and west, and nearly always the upright variety. Persistent cold winds tend to brown the foliage, and prolonged very low temperatures kill the trees.

In the USA, where this cypress was introduced in colonial times, a variety 'Stricta', often only 60 or 90cm (2 or 3ft) in diameter but 6m (20ft) or more tall, is much planted in formal gardens in the warmer regions, especially California; and for this it has no equal among the native cypress species.

Himalayan Cypress

Cupressus torulosa

The Himalayan Cypress is a large, evergreen tree with a bole clad in thick, brown bark which peels away in narrow, fibrous strips. It supports many horizontal or ascending branches bearing the yellow-green scale-like leaves, typical of cypresses. They are very small and incurved at the tips, to make a dense round-topped crown of thread-like branchlets. (See illustration facing p.48.)

Distribution and habitat

The natural range of the Himalayan Cypress extends along the Himalayas from Chamba in Himal Pradesh in the west, as far as the Aka Hills in the North East Frontier area; and then on into western China, in particular the valley of the Upper Yangtse in Szechuan. It is usually rather local in its distribution over the range and at altitudes between 1,830 and 2,750m (6,000 and 9,000ft). It prefers limestone areas, and this factor, combined with its great sensitivity to fire damage, may account for its patchy distribution. At higher altitudes it is frequently found on steep, rocky ground where it is usually growing almost pure. In more favourable situations it is often associated with Deodar *Cedrus deodara*, evergreen oaks *Quercus incana* and *Q. dilitata* and sometimes with Walnut *Juglans regia*. In the far east of its Indian range, in the Aka Hills, it grows with Bhutan Pine *Pinus wallichiana*, against which it is slowly losing ground as the pine is better able to resist fire damage. In China it is a component of the warm temperate coniferous forest together with Japanese Cedar *Cryptomeria japonica*, Chinese Fir *Cunninghamia lanceolata* and other species.

Life history

The round female cones are initially green in colour, flecked with white and purple and growing to about 12mm (0.5in) in diameter as they ripen in the second autumn, to open and release their broad winged seeds. Growth appears to be rather slow but these cypresses eventually reach impressive sizes in favourable conditions. In Garhwal a specimen 47m (154ft) tall, with half its height free of branches, and 6.7m (22ft) in girth, was recorded; and even greater girths – up to 11m (37ft) – have been reported. But over much of the range, especially at the higher altitudes 21 to 24m (70 to 80ft) is more usual with branches emerging much lower down the trunk.

Timber

The light brown, moderately hard, close-grained and fragrant timber is more durable than that of the Deodar but it is much less plentiful. Traditionally used for railway sleepers, it is still used, where available, as building material.

Introduction, performance and use as an exotic

The Himalayan Cypress was introduced into Britain by seeds sent home from India by Dr Wallich in 1824, with a further batch in 1830. In 1838, Loudon saw a 10-year-old plant 3m (10ft) tall at the Horticultural Society's gardens at Chiswick, but he records no others. Elwes says that as many as 34 of the known specimens were killed by the severe winter of 1860-1. A tree planted after this, in 1866 at Hewell Grange, Worcs., was measured by Elwes as being 17m × 1.2m (56ft × 3ft 11in) in 1909. Mitchell found it to be 23m × 1.5m (76ft × 5ft 4in) in 1963 and 23.4m × 1.6m (77ft × 5ft 9in) in 1974. One at Woburn which was 9m (30ft) tall in Elwes' time, had grown to 20.7m × 1.9m (68ft × 6ft 8in) by 1970.

The tallest in Britain in 1971 was at Nettlecombe, Somerset. It measured 27.4m × 2.8m (90ft × 9ft 4in). By 1989 it had grown to 31m × 4.4m (102ft × 14ft 9in).

Other specimens include:-

Abbeyleix, Co. Laois, 21.3m × 1.9m (70ft × 6ft 7in) 1985
Fota, Co. Cork, 25.9m × 2.8m (85ft × 9ft 7in) 1987

Keir House, Perth, 21.9m × 2.4m (72ft × 8ft) 1985
Powerscourt, Co. Wicklow, 28m × 2.2m (92ft × 7ft 7in) 1989
Rossie Priory, Perths., 21.3 × 2m (70ft × 7ft 3in) 1987

This cypress seems to need warm summers to grow satisfactorily, and nearly all the older and larger specimens are in the south or south west of England. Even so, the larger trees tend to succumb to very low temperatures more readily than smaller ones.

In warmer climates the Himalayan Cypress has been much planted as an ornamental. In Australia and South Africa it has proved valuable as a shelter-belt and wind-break tree. In Kenya it has proved more drought-resistant as a forest tree than *C. macrocarpa*; the same has been the case in Zimbabwe where it is thought to have a future, although being very susceptible to fire damage. In India it has been planted as a forest tree in areas with calcareous soils and also as an ornamental in many places on the Plains. It has not, it seems, been planted much in the USA.

Chinese Juniper

Juniperus chinensis

The Chinese Juniper is a conifer with a very wide natural range in China, Mongolia and Japan, and is thus very variable in shape and size. At best it is a forest tree some 18m (60ft) tall with a domed crown of blue-green, blunt, scale-like leaves. It often has a clean, straight bole covered in brown bark which peels off in strips. When open grown, it is pyramidal in shape, with ascending branches emerging from quite low down the bole. At worst it can be no more than a low shrub in areas such as the dry deserts of north east Asia. (See illustration facing p.48.)

Distribution and habitat

In Japan the natural range of the Chinese Juniper is in scattered areas along the Pacific Coast of Honshu, and here and there in Shikoku and Kyushu, where it is found as a rather straggly tree growing in association with Japanese Red Pine *Pinus densiflora*, and various oaks such as *Quercus serrata* and other hardwoods. In China it grows wild in the mountains of the provinces of Kansu, Shensi, Szechuan and Hupeh where it is found mainly as an understorey in the forest with such trees as Walnut *Juglans regia*, Red Cedar *Thuya orientalis* and both evergreen and deciduous oaks. It also grows as a shrub on barren sand dunes in the Mongolian deserts.

In both China and Japan this juniper is a very commonly planted tree, particularly around temples and tombs, where there are some ancient specimens, and in parks and gardens.

Life history

The flowers appear in April, the males and females often on separate trees. The females develop into succulent, berry-like cones, green with a glaucous bloom in the first year, but turning a purplish brown when ripe in the second autumn.

Seedlings grow quickly and young plants often put on 45cm (18in) a year, but growth eventually slows up. Mature trees are generally about 15m (50ft) tall, but some very old ones have relatively huge girths. Elwes saw some at Confucius' tomb which were reputed to be 700 years old and, although only some 12m (40ft) tall, were up to 5m (17ft) in girth. Others in Japan were much taller – 21 to 24m (70 to 80ft) – and though larger in girth too, their trunks were hollow. At a temple near Osaka there was a tree with a 3m (10ft) clear trunk, 3m (10ft) in girth, with a branch spread of over 12m (40ft).

Timber

The reddish-brown timber is not in great supply. It is used for making pencils, for carving and sometimes for poles and posts.

Introduction, performance and use as an exotic

The Chinese Juniper was first described by Linnaeus in 1767 when it was already under cultivation in Upsala, Sweden. It first came to Britain when William Kerr sent plants home from Canton in 1804.

It does not seem to be very long-lived in Britain. Mitchell could only find two of the six trees mentioned by Elwes. One of these at Eastnor Castle, Herefordshire, which was 14.6m × 0.9m (48ft × 3ft) in 1908, and bearing fruit, had grown to 15.8m × 2.4m (52ft × 7ft 11in) by 1969; and the specimen at Westonbirt, Glos., which was 9.7m (32ft) tall in 1909 and covered with cones, had grown to 17.9 × 1.5m (59ft × 5ft) by 1970; and 10 years later it had grown to 20.7m × 1.6m (68ft × 5ft 4in) making it possibly the tallest in Britain.

Other recent measurements include the following:-

Bedgebury Pinetum, Kent, 17m × 1.4m (56ft × 4ft 10in) 1981
Bicton, Devon, 18.9m × 1.7m (62ft × 5ft 9in) 1977
Caledon Castle, Co. Tyrone, 18.3m × 2.2m (60ft × 7ft 6in) 1985
Castlehill, Devon, 18.9m × 1.8m (62ft × 6ft) 1989

On the whole, the hardy Chinese Juniper is rather slow-growing in Britain which is perhaps no bad thing when it is planted in a small garden. Apart from the type tree, there are several cultivars which are commonly planted, one of which is 'Aurea'. It occurred at Milford, Surrey, in 1855 and was distributed in 1872. It is a golden tree, compact, always male and capable of growing to 12m (40ft) tall. Another is 'Keteleeri', a dark green, narrowly conic tree about 9m (30ft) tall which appeared in Belgium around 1910.

The Chinese Juniper was introduced into the USA in 1767 and, although very similar in appearance and hardiness to the native *J. virginiana*, it has proved very variable and some 10 or so cultivars and varieties are offered by the trade, mostly of narrow and compact growth. These include 'Mountbatten', introduced by Sheridan Nurseries, Toronto, a densely compact, grey-green tree giving the impression that it has been sheared; 'Story', a very narrow, columnar tree selected by Professor T.J. Maney of Iowa State College in 1935 and introduced in 1947, and a variety *mas*, again densely columnar.

Chinese Thuya

Thuya orientalis

The Chinese Thuya is a much smaller tree than the American species, often with several stems emerging from the base. The scale-like leaves are borne in vertical plates and are the same dark green on both sides. In old age the trees become straggly with broad, uneven crowns, but when young they are often narrowly conic and well furnished, with branches down to ground level. (See illustration facing p.49.)

Distribution and habitat

The Chinese Thuya is another tree so widely planted in China, its native country, that its natural range is not easy to define, but it was probably confined to the mountains of north east China from Hopei in the north to Shensi in the south. Here it grows mostly at around 1,525m (5,000ft) on steep, rocky, arid slopes in company with Chinese Pine *Pinus tabulaeformis* and Temple Juniper *Juniperus rigida*, as well as on the more fertile valley slopes with Evergreen Oaks *Quercus dentata* and *Q. spinosa*, and other deciduous trees such as chestnuts and maples.

This thuya is also very widely planted in Japan, and has been for centuries, so much so that early travellers thought it indigenous to that country where it grows in temples, grave-yards and private gardens.

Life history

The female cones, which are round and much more cypress-like than the urn-shaped ones of the American species, are fleshy with a bluish bloom at first, but turn brown and woody as they ripen in the same autumn, when the hooked scales open to release the large, unwinged seeds.

Growth is slow and the trees rarely exceed 18m (60ft) in height and usually much less in the relatively hostile sites where it still grows wild, often forking with several stems. The trees planted in sheltered places are those which attain reasonable sizes.

Timber

It is reported that the timber was used for making coffins for Chinese princes but, being mainly an ornamental tree when of any size, its timber is of little or no commercial value.

Introduction, performance and use as an exotic

The Chinese Thuya was growing in Europe, at Leyden, some time before 1737 when Linneaus described it. These early plantings may have been grown from seed sent to Holland from Japan by Kaempfer who travelled to that country in 1690-91. Seeds were also sent home to Paris around that time by French missionaries working in northern China. The first mention of this tree in Britain was in a letter from the Duke of Richmond to Collinson, dated 1 February 1743, in which he bemoaned the fact that Miller would not give him one of the first specimens to arrive in the country. But Loudon gives the introduction date as 1752, and quotes details of a number of specimens between 20 and 60 years old in 1838, most of them about 10.5m (35ft) tall. None of these seems to have been found by Henry who mentions a pair at Hampton Court, Hereford, which were 12m (40ft) tall in 1906, but Gardner could not find them in 1945. So, perhaps this thuya is not long-lived in Britain. Mitchell records details of a tree at the West Suffolk Hospital, Thetford, which was planted in 1865 and measured 8.5m × 1.3m (28ft × 4ft 7in) a hundred years later; and another at Kew which was planted in 1898 and was 12.8m × 1.3m (42ft × 4ft 7in) in 1965. The two tallest trees in the country, planting date unknown, were at Penjerrick and Pencarrow, Cornwall, and measured 15.8m × 1.5m (52ft × 5ft) and 15.8m × 1.8m (52ft × 6ft 4in) in 1959 and 1970 respectively. Some more recent measurements are:-

Battersea Park, London, 14.6m × 0.7m (48ft × 2ft 5in) 1979
Culverwood, Herts., 14m × 1.5m (46ft × 5ft) 1985
East Malling Church, Kent., 15.2m × 1.6m (50ft × 5ft 7in); 12.3m × 1.2m (42ft × 4ft 2in) 1984
Ewhurst Church, Sussex, 12.2m × 1.5m (40ft × 5ft 2in) 1986

Mitchell says that the Chinese Thuya is common in small gardens, but is usually unhealthy, the best specimens being in towns and cities and on well-drained, even hungry, soil. A very attractive cultivar, 'Elegantissima', of an upright, flame-like shape when young, and tinged a bright golden colour in the summer, is widely planted in Britain.

This thuya was introduced into the USA some time before 1737 and a number of cultivars and garden varieties are extensively planted, including 'Elegantissima' and 'Baker' which has bright green foliage and a considerable ability to withstand hot, dry conditions.

In parts of India and in Hong Kong the Chinese Thuya is planted as an ornamental, and a particularly upright form is frequently grown in tubs as well as directly in the ground.

White Cedar

Thuya occidentalis

The White Cedar is a tree of the northern latitudes of eastern North America, forming dense, pure stands on cool, swampy ground and along the rocky beds of streams. Never so large as the Western Red Cedar, it otherwise resembles it greatly, with the same buttressed trunk clad in red-brown, shredding bark, but with a more open crown of rather yellowish-green, scale-like leaves. (See illustration facing p.49.)

Distribution and habitat

The natural range of the White Cedar extends from the Gulf of St Lawrence in Canada westwards to south east Manitoba, and from New England and New Jersey west to central Minnesota, around the Great Lakes to Vermont and New Hampshire; and in isolated areas down the Appalachians as far south as east Tennessee and western North Carolina.

This range includes the extremes of the tundra boreal forests of Canada and the relatively warm climate of the South Appalachians, although this is tempered by altitudes of around 920m (3,000ft).

The White Cedar fares best on swampy sites and on soils which are neutral or of limestone origin, when stands are often pure and dense. In upland areas it is found in meadows and on shallow, rocky pastures.

Common associates of the White Cedar are, in the north, Aspen *Populus tremuloides*, Tamarack *Larix laricina*, White Spruce *Picea glauca*, Balsam Fir *Abies balsamea*, and Black Spruce *Picea mariana*; and, in the south, Black Ash *Fraxinus nigra*, Red Oak *Quercus rubra* and Sugar Maple *Acer sachharum*.

Life history

The urn-shaped cones are produced in small quantities at the age of six years, and in profusion at 20 years and after. They ripen in October and the seeds are dispersed by wind, to germinate in the following May and June. The seedlings grow strong tap roots in early years.

The White Cedar is never a large tree, the average height in the north being 12.2 to 15.2m (40 to 50ft), with girths of 2 to 3m (6 to 9ft); and on upland sites in the south, where growth is quicker 24m (80ft) may be reached. But even on the best sites it may take as long as 160 years to reach this sort of size. However, the tree is long-lived, attaining ages of up to 400 years; and as it can readily stand shade from associated taller species and responds well to release, it is not easily ousted from mixed stands.

Timber

Although not available in the large dimensions of Western Red Cedar, the timber of White Cedar is similar in lightness, durability and ease of working, as well as in colour and appearance. It is used for many of the same purposes including shingles, posts, poles, boats and other light work where durability is essential.

Introduction, performance and use as an exotic

The White Cedar was probably the first American conifer to be introduced into Europe. It appeared in France during the reign of Francis I around 1500, as there was reportedly a specimen growing at Fontainebleau at that time, sent to the king as a gift from Canada. Gerard, writing in 1596, said that the White Cedar grew in his English garden in profusion.

None of the specimens listed by Loudon exceeded 9m (30ft) in height, and those measured by Elwes were never more than 13m (42ft); and none listed by either man has been found by Mitchell, suggesting that the tree is not long-lived in Britain. Indeed, by 1970, Mitchell had found only two good specimens in the whole country; one at Carey House, Dorset, which was 20.7m × 1.8m (68ft × 5ft 11in) in 1968, and the other at Trawscoed, Brecon, which was 19.2m × 2m (63ft × 7ft 2in) in 1969. To these he has recently added four more:

Bishopthorpe, York, 12.8m × 1.5m (42ft × 5ft 2in) 1989
Little Hall, Kent, 18.9m × 1.3m (62ft × 4ft 5in) 1979
Logie House, Nairn, 12.5m × 1.9m (41ft × 6ft 7in) 1981
Lydney Church, Glos., 14m × 1.3m (46ft × 4ft 5in) 1983

The other old trees which he found were thin and straggly. Clearly, the tree does not thrive in Britain. Even young specimens are inclined to be spoilt by an excessive production of cones, rendering them unsightly.

The species itself is thus not much planted but, like the Lawson Cypress, and perhaps only second to it, the White Cedar has produced a great many cultivars, most of which are small or dwarf and are freely planted by gardeners. The larger cultivars include 'Lutea', which grows strongly up to 15m (50ft) tall, with branch-ends touched with yellow; 'Spiralis', narrow and columnar, up to 12m (40ft); and 'Vervaeneana', conic, yellowish and up to 9m (30ft) tall.

In the USA this native thuya has produced many slow-growing forms which do well in areas where there is considerable atmospheric moisture. These include 'Douglasi Pyramidalis' which is dense and broadly columnar; *fastigiata*, narrow and columnar; and 'Lutea' which originated in Geneva, New York, around 1873 and grows tall and pyramidal, and, unlike most others, retains its full colour throughout the growing season. In fact, a failing of the garden varieties of this species, wherever they are grown, is their tendency to go brown in the winter, just when their colours are most needed.

Western Red Cedar

Thuya plicata

The Western Red Cedar is one of the giants of the great Pacific Coast forests of north west America and, next to the Douglas Fir, economically the most important species. The huge buttressed trunk, with a cinnamon-red, fibrous bark, peeling in strips, carries short, horizontal branches. These bear a narrow, pyramidal crown of dark, glossy green, scale-like leaves, which overlap one another, very like the cypresses. Even in close-grown stands the lower branches on younger trees tend to be retained for up to 50 years, but old trees often have boles free of branches for up to half their 60m (200ft) height.

Distribution and habitat

The natural range of the Western Red Cedar extends from the coastal region of southern Alaska south through the coastal ranges of British Columbia, western Washington and Oregon, down to Humbolt County in California. In British Columbia it extends east to the western slopes of the Continental Divide and south from there to the Salmon River Mountains of Idaho. The eastern limit is the western slopes of the Rockies in Montana.

This species is confined to regions which have abundant rainfall, and it reaches its maximum development in the coastal areas on moist flats and along river banks. But it is also found at varying elevations up to 2,140m (7,000ft), although the trees are smaller in size as altitude increases. Those growing higher up than 920m (3,000ft) are rarely of any economic importance.

The Western Red Cedar's main associates, for it is not often found growing in pure stands, are Western Hemlock *Tsuga heterophylla*, Douglas Fir *Pseudotsuga menziesii*, Sitka Spruce *Picea sitchensis*, Grand Fir *Abies grandis*, Pacific Silver Fir *Abies amabilis*, Coastal Redwood *Sequoia sempervirens*, Big-leaf Maple *Acer macrophyllum* and Red Alder *Alnus rubra*, in the coastal areas; and higher up, Mountain Hemlock *Tsuga mertensiana*, Noble Fir *Abies procera*, Western White Pine *Pinus monticola*, and Subalpine Fir *Abies lasiocarpa*.

Life history

The little urn-shaped cones, borne sparsely on trees as young as 16 years, but in great profusions on older trees every two or three years, start to release their tiny winged seeds in September and October. They continue to do so gradually throughout the winter. Germination then takes place in the autumn in coastal areas and the following spring in the mountain regions.

Compared with most of its competitors, the Red Cedar is rather slow growing, increasing in height by no more than 60cm (2ft) annually at first, then slowing down to attain a height of about 30m (100ft) in 80 years. Under favourable conditions, such as in the Puget Sound area of Washington, it reaches enormous sizes – more than 60m (200ft) tall with girths of up to 12m (40ft); but in mountain areas, although the height may be as much as 53m (175ft), the girths rarely exceed 7.5m (25ft). Some of the largest trees are thought to be 800 to 1,000 years old, but the centres of their boles are always hollow.

The Red Cedar is exceptionally shade-tolerant, growing well and even reaching maturity in the shade. When competition is removed response is immediate. Heavy seedling mortality may, however, allow hemlocks and firs to gain a permanent dominance in young stands.

Timber

The aromatic, red-brown timber is medium grained, soft, light and rather brittle. But it is easily worked, scarcely warps or shrinks and has a great resistance to decay and insect attack. These virtues make it a very valuable timber for the exterior of buildings, for shingles, boat building, poles, posts, greenhouses and the many other uses for which it is in great demand.

Introduction, performance and use as an exotic

The Western Red Cedar was originally discovered by Nee who accompanied Malaspina on his voyage round the world in 1789-94. The original specimen, gathered at Nootka Sound, is now in the Natural History Museum, South Kensington. Archibald Menzies also gathered specimens at the same place in 1795, but it was not until 1853 that the tree was introduced to cultivation in Britain when Lobb sent seeds to Veitch's nursery.

One of the trees grown from these original seeds is still growing at Dropmore. It was 20.7m × 1.8m (68ft × 6ft) when Elwes measured it in 1905, and 24.4m × 2.7m (80ft × 9ft) in 1970 when Mitchell measured it and noted that rooted layers round the main trunk were giving it serious competition and stunting its growth. Another specimen at Stourhead, Wilts., planted in 1854, had grown to 37.8m × 4.8m (124ft × 15ft 11in) by 1970.

The finest Red Cedar which Elwes found and photographed was a tree at Fontill Abbey, Wilts., grown from seed sown in 1860. It measured 28.9m × 3m (95ft × 10ft) in 1906. It had grown to 35m × 4.2m (115ft × 14ft 3in) when Gardner measured it in 1947, and was 36.5m × 4.5m (120ft × 14ft 11in) when Mitchell saw it in 1963.

This thuya grows rapidly in Britain – 24m (80ft) in 40 years on good sites – with girth increases averaging 5cm (2in) a year for up to 60 years. The biggest trees (mostly in the north and west) of the many fine specimens in parks and gardens are about 41m (135ft) tall with girths anything up to 4.5m (15ft) and more. Recent measurements include the following trees:-

Balmacaan, Inverness, 41m × 6m (135ft × 19ft 11in) 1987
Belladrum, Inverness, 41m × 5.7m (135ft × 19ft 3in) 1987
Castlehill, Devon, 42m × 5.4m (138ft × 18ft) 1989
Powerscourt, Co. Wicklow, 42m × 5.8m (138ft × 19ft 8in) 1987
Stourhead, Wilts., 35m × 5m (115ft × 17ft) 1987
Strone House, Argyll, 45.7m × 5.4m (150ft × 18ft) 1989

An attractive cultivar 'Zebrina', which originated in Holland in the 1860's, with the foliage banded in greenish-yellow and gold, is also often planted, the largest of these being 18 to 24m (60 to 80ft) tall.

As a forest tree, the Western Red Cedar is becoming increasingly popular. Its growth rate compares very favourably with Douglas Fir, Western Hemlock and Sitka Spruce. It is tolerant of a wide range of soil types from sandy loams and heavy clays to limestones and chalk; and being a shade-bearer it is particularly useful for underplanting in native broadleaved woodland and under larch. Furthermore, the timber grown in Britain is at least as strong as that grown in America and perhaps as durable. When larger sizes are available it will have a good market.

It produces viable seed in Britain and several seed stands have been registered. But as far as nursery cultivation is concerned, the susceptibility of seedlings to the fatal fungus *Keithia thuyina* is a serious drawback.

In the USA this native thuya is rated more satisfactory for garden use than *Th. occidentalis* in northern areas because it does not turn brown in the winter. Plants grown from seed collected in the high mountains of Montana and Utah prove to be hardy in the north eastern states.

A variety, *atrovirens*, with very dark green foliage, is also often planted in the United States.

Hiba

Thuyopsis dolobrata

The Hiba is a conifer exclusively native to Japan. It is cypress-like at first glance but with bigger and more striking, bright green, scale-like leaves, very intensely white on the undersides. These leaves are borne on slender branches which spread and droop to form a rounded, conical crown. The reddish-brown bark peels off in strips, leaving grey scars. (See illustration facing p.64.)

Distribution and habitat

The natural range of the Hiba stretches from the southern tip of Hokkaido in the north right down through the country to Kyushu in the south. It occurs mainly at altitudes between 920 and 1,830m (3,000 and 6,000ft), usually in mixture with other species, but in the extreme north of Honshu and, in particular, on the peninsular of Shimokita and Tsugaru some of the best known forests in Japan are of almost pure Hiba. Elsewhere, it is associated with such trees as Red Cedar *Thuya standishii*, White Pine *Pinus parviflora* and Larch *Larix kaempferi*.

Life history

The flowers appear in April and May, the females, at the ends of the shoots, developing into small, round, blue-grey cones which ripen brown and woody, with thick scales from which they cast their seeds in the same autumn.

Seedling growth is slow, but trees in the forest reach 21 to 24m (70 to 80ft) in height in 150 to 200 years, with girths of 1.8m (6ft) or so; some in sheltered spots may even attain 30m (100ft).

This tree is extensively planted in Japan as an ornamental.

Timber

The yellowish-white, fine grained timber is very durable and is greatly valued in Japan for construction work; having an aromatic scent and taking polish well, it is used for furniture-making, for carving and as a base for lacquer work.

Introduction, performance and use as an exotic

The Hiba was first introduced into Britain by T. Lobb, Veitch's collector, who sent a single plant from the Botanic Gardens, Buitenzorg, Java, to the Exeter nursery in 1853, but it died. Another plant, sent home by Earl Fortescue's brother from Japan, and planted at Castlehill, Kent, in 1859, also died, but not before cuttings had been taken from it. In 1861, both Veitch and Fortune sent home seeds from Japan and from these a general distribution was made.

The finest tree which Elwes saw was at Killerton, Devon, which measured 10.6m × 0.7m (35ft × 2ft 4in) in 1902. By 1970 it had grown to 16.7m × 1m (55ft × 3ft 5in) and, by layering, had developed 17 stems in a circle 13.7m (45ft) in diameter.

The Hiba is common in many gardens as a very attractive shrub, but in the west of Britain, and particularly in the south west, reasonably sized trees flourish. Mitchell listed 41 specimens in 1970 of which the tallest were two trees at Tregrehan and Penjerrick in Cornwall, which both topped 21m (69ft). By 1987 the Tregrehan tree had grown to 21.3m × 2.6m (70ft × 8ft 10in). Other measurements made recently include:-

Boconnoc, Cornwall, 22.8m × 1.6m (75ft × 5ft 6in) 1983
Glendurgan, Cornwall, 21.9m × 1.6m (72ft × 5ft 4in) 1984
Portmeirion, Gwynedd, 23.4m × 1.2m (77ft × 4ft 2in) 1986
Sheffield Park, Sussex, 21.3m × 1m (70ft × 3ft 8in) 1989
Woodstock, Co. Kilkenny, 25m × 1.5m (82ft × 5ft 1in) 1989

The Hiba does not set seeds in Britain, but it is very easily propagated by cuttings. A rather unstable cultivar 'Variegata', with whole shoots growing white here and there, is often found in gardens.

This tree was introduced into the USA in 1861 but does not seem to be much planted; Wyman says 'It is of no particular merit in preference to the arborvitaes'.

Redwood

Sequoia sempervirens

The Redwoods of California are perhaps the world's most famous and best known trees. They are the tallest in the world and the quickest growing conifers in North America. Their immensely tall trunks are straight, slightly tapered and heavily buttressed and often clear of branches for 45m (150ft). The bark is soft and spongy and the trunk support crowns of short, horizontal branches bearing shiny green needles which are set comb-like along the shoots. Standing in great cathedral-like groves, they are one of the most impressive sights in nature. (See illustration facing p.64.)

Distribution and habitat

The natural range of the Redwoods extends in a narrow coastal strip, 720 kilometres (450 miles) long and 5 to 56 kilometres (3 to 35 miles) wide, from Checkto River in the extreme south west of Oregon to Salmon River Canyon in the Santa Lucia Mountains of south Monterey County in California. This range is limited by and corresponds to the summer fog belt which prevents high temperatures and maintains a humid atmosphere during hot summer days.

The best stands are found on the coastal flats in deep sedimental soils and on the gentle slopes of valleys opening towards the sea. But they also grow as high up as 920m (3,000ft) although the main body of them is found between 30 and 760m (100 and 2,500ft).

Pure stands are found only on the best sites, the proportion of associated tree species increasing with altitude. The most numerous associate is Douglas Fir *Pseudotsuga menziesii* which is found throughout the range: others are more limited in their distribution, such as Western Hemlock *Tsuga heterophylla*, Grand Fir *Abies grandis*, Sitka Spruce *Picea sitchensis*, being fairly common; with Lawson Cypress *Chamaecyparis lawsoniana*, Western Red Cedar *Thuya plicata*, Madrone *Arbutus menziesii*, Big-leaf Maple *Acer macrophyllum* and Vine Maple *A. circinatum* less so.

Life history

Prolific seed bearing starts at about 20 years of age and optimum production is reached at 100 years, with good seed years almost annually. The flowers bloom between November and March, the cones maturing and dropping their seed in the following autumn. The small proportion which germinate do so immediately and the seedlings grow very quickly – 45cm (18in) in the first year – and maintain this initial impetus for many years.

The height growth of young trees is spectacular, many reaching 30 to 45m (100 to 150ft) in 50 years, and 49.4 to 66m (165 to 220ft) in 100 years. The growth in girth is no less impressive, being as much as 7.2m (24ft) in 100 years.

The tallest Redwood so far found is 112m (368ft) (a twin-stemmed tree) and many more are well over 91m (300ft). The oldest so far measured by accurate ring counts was 2,200 years, but most of the mature trees in the old stands are probably between 400 and 500 years old.

The Redwood is very shade-tolerant, growing in dense stands. Young trees can endure intense shade almost indefinitely (one 25cm (10in) tree was found to be 100 years old). It always forms climax forest and is always dominant when growing with other trees. The affection in which the Redwoods are held, because of their immense size, great age and surpassing beauty, and the damage done to virgin stands by logging operations, has resulted in a number of Redwood Parks being established in the 'Redwood Empire', with money raised by charitable bodies, so these natural stands can be protected from further exploitation.

Timber

The timber is reddish-brown, straight-grained, soft, moderately strong and very durable. It has many uses in which large sizes are required, such as general construction work, bridge timbers and railway sleepers, as well as for shingles, furniture, wooden pipes, flumes and much else.

Introduction, performance and use as an exotic

The Redwoods were discovered in 1769 by Father Crespi, the recorder of a Spanish expedition, on low hills round Monterey Bay. Archibald Menzies was the next to report on them in 1795, followed by David Douglas in 1831. But they were probably first introduced into Britain by means of seed sent by Dr Fischer of the St Petersburg Botanic Gardens to Knight and Perry's Chelsea Nursery in 1843 (the Russians had a settlement at Fort Ross, California at that time). This date, for which there is no definite confirmation, seems reasonable; a tree, 30cm (1ft) tall, was

Hiba, Polesdon Lacey, Surrey (p.61)

Redwood, Tall Tree Grove, Calif.
World's tallest tree (r.) 112m/367ft (p.63)

Wellingtonia, Baynard's Park, Sussex (p.66)

Wellingtonia, Mariposa Grove, Yosemite, Calif. (p.66)

Japanese Cedar, Leighton, Welshpool, Powys (p.69)

Japanese Umbrella Pine, Bodnant, N. Wales (p.71)

Japanese Umbrella Pine, Kyloe Wood, Northumberland (p.71)

Chinese Fir, Pencarrow, Cornwall (p.73)

bought for 5 guineas (£5.25) from the same Knight and Perry in 1845 and planted at Dropmore, Bucks. This tree was 28.6m × 3.3m (94ft × 11ft) when Elwes measured it in 1907, and Mitchell found it to be 33m × 4.7m (108ft × 15ft 8in) in 1970. An even older tree, this time of unknown origin, but probably from Knight and Perry, was planted at Smeaton House, East Lothian, in 1844 and was 17.3m × 2.7m (57ft × 9ft) in 1902, according to Elwes, having lost its top several times.

The finest Redwood known to Elwes and which he photographed, was at Claremont. In 1907 it was 29.8m × 3.8m (98ft × 12ft 9in), and in 1965 Mitchell found it to be 36m × 5m (118ft × 17ft 1in).

A Redwood at Cuffnels, New Forest, planted in 1855, was found by Elwes to be 32m × 3.2m (105ft × 10ft 10in) in 1906. Some 65 years later it had grown to 39m × 4.9m (128ft × 16ft 6in); and by 1980 it had grown taller by 3m (10ft) and had put on 30cm (1ft) in girth. Seven other Redwoods, planted in the 1850's and measured by Elwes around 1906, averaged 27m (90ft) in height and 3.3m (11ft) in girth. Some 65 years later Mitchell found them to have attained an average increase in height of 9.6m (32ft) and 1.8m (6ft) in girth. Of 100 or so old trees of which Mitchell gave details in 1970, including those mentioned above, the tallest was one in North Devon which topped 42m (138ft), and the largest was at Taymouth Castle, Perthshire, which girthed 6.7m (22ft 6in).

In 1989 the tallest specimen was at Bodnant, Gwynedd; it measured 46.3m × 4.8m (152ft × 16ft 5in). And the tree at Taymouth Castle measured 39.6m × 7m (130ft × 23ft) in 1983. Other recent measurements are:-

Caledon Castle, Co. Armagh, 37m × 7m (121ft × 23ft 7in) 1985
Eastnor Castle, Herefords., 35m × 6m (115ft × 20ft) 1987
Longleat, Wilts., 32m × 6.5m (105ft × 21ft 6in) 1986
Melbury, Dorset, 37m × 6.8m (121ft × 22ft 7in) 1989
Rossie Priory, Perths., 6.1m × 7.5m (120ft × 20ft 2in) 1985
Taymouth Castle, Perth, 37.8m × 6.7m (124ft × 22ft 9in) 1983
Woodstock, Co. Kilkenny, 42m × 7.6m (138ft × 25ft 4in); 43m × 7.6m (141ft × 25ft 2in) 1989

Redwoods make very attractive ornamental trees in Britain, with their red bark and the symmetry of their dark green crowns, but they grow large very quickly and are not suitable for small gardens.

The first plantation of Redwoods to be established in Britain was at Leighton, near Welshpool, Powys, now owned by the Royal Forestry Society. The 33 trees were brought over from America in pots and planted in 1858. They now average 35m (115ft) in height – the tallest is 41m (135ft) – but seem unlikely to grow much taller because of exposure to wind. The volume of timber in this very beautiful stand is 2150cu. metres to the hectare (31,000 cu.ft to the acre), a record for standing timber in Britain. There is a fine stand, much younger, at Dartington, Devon, planted in 1934. In 1965 the top height of these trees was 26m

(85ft) and the volume 1,040 cu. metres to the hectare (15,000 cu.ft to the acre). In 1983 they averaged 31m (101ft) in height, but the stand was severely damaged by gales in 1990.

The Redwood is very shade-tolerant in Britain and it certainly has a future as a forest tree, both in pure stands and as an underplanting in hardwood crops, at least on favourable sites in south west England and the Welsh Borders.

The timber grown in Britain has not yet been produced in sufficient quantity to enable its value to be assessed.

Wellingtonia or Giant Sequoia

Sequoiadendron giganteum

The Wellingtonia is perhaps the world's most spectacular tree. In volume greater than any other on earth, its immense bole, swollen at the base where it is often 60m (200ft) in circumference, and covered in 60cm (2ft) thick red-brown, spongy bark. The trunk supports a rounded crown of deep green, scaly, sharp-pointed leaves at a height of over 75m (250ft) the whole so massive and awe-inspiring as to be almost beyond belief. Young trees up to 100 years of age retain their downswept branches which reach almost to the ground, whilst the old trees are often bare of branches for up to 45m (150ft). (See illustration facing p. 64)

Distribution and habitat

The Wellingtonia is found growing naturally in 70 more or less isolated groves on the western slopes of the Sierra Nevada in central California, in a belt about 420 kilometres (260 miles) long. The northernmost grove, consisting of only six trees, is along the Middle Fork of the America River in Placer County, and the southernmost, with 100 trees, near Deer Creek in Tulare County. Most of the groves are at altitudes of between 1,370 and 2,290m (4,500 and 7,500ft); some as low as 920m (3,000ft) in the north and as high as 2,670m (8,900ft) in the south. They are usually found in canyons where soil moisture is always adequate, most of it resulting from snow which, in winter, may be up to 6m (20ft) deep.

The most common associated trees are White Fir *Abies concolor*, Sugar Pine *Pinus lambertiana*, Ponderosa Pine *P. ponderosa*, Incense Cedar *Calocedrus decurrens* and Californian Black Oak *Quercus kellogii.*

Life history

The round cones are borne by trees as young as 18 to 25 years but seeds of these are normally infertile; only after they reach 150 to 200 years do they start to produce large quantities of viable seed. Later on still, the biggest trees bear enormous crops of cones; it is not unusual to find up to 2,000 in one year on a single tree.

Despite the huge quantities of seed which fall and are dispersed by wind, very few actually germinate, a proportion being destroyed by birds and squirrels. Of the seedlings of all species which appear in a grove only one per cent are Wellingtonias. Those which do appear grow up to 7cm (3in) during the first year, and by the end of the third are often 30cm (1ft) tall, and well able to compete with their rivals.

In the sapling stage they grow rapidly, expanding the girth of their potentially massive trunks by several centimetres every year until in the oldest specimens, which may be well over 3,000 years of age, timber volumes may be as much as 1,175 cu. metres (42,000 cu.ft). One measured tree contained 1,400cu. metres (50,000cu.ft) of timber. The tallest Wellingtonia at present is the MacKinley Tree, in one of the southern groves; it stands at 88.7m (291ft). Top heights of 84m (275ft) are common.

After a visit to the Mariposa Grove in 1979, I wrote in *Quarterly Journal of Forestry*:-

'To stand beside one (a Wellingtonia), dwarfed by its gigantic trunk on a buttressed base 30m (100ft) in diameter, is to be in the presence of something which has achieved its uniquely massive status in the botanical world by surviving millennia of pre-history, and yet still supports a huge green crown studded with cones which still scatter viable seeds. One can but marvel at such tenacity of life in the face of a climate which must have changed considerably, and in the inferno of forest fires which must have swept by during the eons of time of the trees' witness. An apparent immunity to insect and fungi attack, a prodigiously thick and fire-proof bark and a massive, wind-firm base have ensured this survival. Even fire, its greatest enemy, can also be its greatest friend, for it leaves it standing amidst the wrecks of competing pines and firs, with the ground beneath it clear enough of undergrowth to give its seedlings a chance to prosper.'

Young trees bear shade better than their elders, but the latter, whilst not competing very well for light with their associates, are better able to withstand drought than most of them, and in this respect have an advantage, especially over the most vigorous competitor, the White Fir *Abies concolor*.

Timber

The timber is dull purplish-red and rather brittle; but the huge logs are so inaccessible and so difficult to handle that they have no commercial value.

Introduction, performance and use as an exotic

The Giant Sequoia was discovered by John Bidwell in 1841 in what is now the Calaveros Grove. J.D. Matthews was the first to send a small consignment of seeds to Britain from this same grove in 1853, the year the Duke of Wellington died, hence the name. But Lobb returned home in the same year bearing large quantities of seed (and two small plants) again from the Calaveros Grove. From this consignment most of the original trees in Britain were grown.

The tree rapidly became immensely popular and specimens were planted in almost every estate policy and large garden throughout the realm. Being resistant to most diseases and tenaciously wind-firm, most of them are still flourishing, except those which have been struck by lightning or felled for development.

One of particular historical interest, planted at Stratfield Saye, Berks., the old Duke of Wellington's seat, in 1857, measured 26m × 3.6m (85ft × 12ft) when Elwes saw it in 1903; and it was 27m (90ft) tall four years later. In 1968 Mitchell found it to be 36m × 7.3m (118ft × 24ft 4in).

A specimen from Matthews' seed planted out at Glendoik, Scotland, in 1860, was 25.3m × 5m (83ft × 17ft 3in) in 1931 and 36m × 5.8m (118ft × 19ft 6in) in 1970.

Elwes provided a photograph of a fine tree at Fonthill Abbey, Wilts. which, in 1906, measured 32m × 5m (105ft × 17ft). In 1957 Gardner found it to be 40.3m × 7.3m (165ft × 24ft). A tree at Endsleigh, Devon is the same height – about as much as many of the Wellingtonias have so far managed.

Some much younger trees planted in the early years of the 20th century are already 24m (80ft) and more tall, but like the older ones will probably be halted in their height growth by wind.

Recent measurements include the following trees:-

Altyre, Morays, 43m × 8.5m (141ft × 28ft 3in) 1987
Castlehead, Devon, 52.4m × 8.8m (172ft × 29ft) 1988
Murthly Castle, Perth, 45m × 7.3m (148ft × 24ft 5in) 1987
Penrhyn Castle, Gwynedd, 42m × 8.3m (138ft × 27ft 7in) 1987
Powerscourt, Co. Wicklow, 39.6m × 8.8m (130ft × 29ft 2in) 1988
Powis Castle, 40.2m × 9.7m (132ft × 32ft 3in) 1987
Scone Palace, Perth, 41m × 9m (135ft × 30ft) 1987

Few exotic trees have made a greater impact on the English countryside than the Wellingtonias. Wherever there are estate policies or large gardens these magnificent trees can be seen towering above the broadleaved species which surround them; and often individual trees stand in fields, massive and stately as though they have been there since the beginning of time.

Few forest plantations of Wellingtonias have ever been established. One at Aston Clinton, Bucks., planted in 1869, grew very well and had a top height of 21m (70ft) in 1906, but unfortunately it was felled a few years ago. A stand in Kyloe Wood, Northumberland, planted in 1902, consists of 67 trees with an average top height of 34m (114ft) and a total volume of 446cu. metres (15,777cu.ft).

It is surprising to find that this tree, native to high altitudes and extremes of cold in the Sierras does not do well in the north eastern areas of the United States. A 18m (60ft) specimen near Cayuga Lake in central New York was killed outright by the severe winter weather of 1933-34; and they never survive for any length of time in the Arnold Arboretum, Washington D.C.

Japanese Cedar

Cryptomeria japonica

The Japanese Cedar is a huge tree with a massive straight bole, tapering from a broad base above the roots. The reddish-brown bark is soft and fibrous and peels off, hanging in long strips. The trunk supports a conic crown, with a rounded top of branches which droop lower down. The leaves are upswept and awl-shaped, curving inwards, not unlike those of the Wellingtonia. (See illustration facing p.65.)

Distribution and habitat

The natural range of the Japanese Cedar in China is probably confined to the mountains of Chekiang and Fukien Provinces where it is a component of the warm temperate coniferous forests in company with such trees as the Chinese Fir *Cunninghamia lanceolata*, Chinese Hemlock *Tsuga chinensis*, and Golden Larch *Pseudolarix amabalis*. It has, however, been so extensively planted as a forest and ornamental tree in other provinces, like Kiangsi, Hupeh, Szechwan and Yunnan, where it has become naturalised, that it is difficult to define accurately the areas where it originated.

In Japan it is found in the whole of Honshu, except the northern tip, throughout Shikoku and in scattered areas in Kyushu. In the south it grows at altitudes up to 1,830m (6,000ft) and in the north up to 610m (2,000ft), but the best stands are between 300 and 1,070m (1,000 and 3,500ft). They grow almost pure on the north facing slopes but in mixture with broadleaves such as Japanese Horse Chestnut *Aesculus turbinata, Acer rufinerve, Magnolia* spp. and *Quercus* spp. on south facing aspects. As in China, this tree has been planted everywhere in Japan for centuries, both as an ornamental and a forest tree. There are some magnificent specimens near temples and in avenues along approaches to towns, which are as huge and impressive as they are immensely old.

Life history

The flowers appear in March and April, the females at the ends of the branchlets, ripening in the second October and November and remaining on the tree after they have shed their seeds.

Growth is quick and saplings may be 0.9 to 1.2m (3 to 4ft) tall after only two years. Coppice shoots also grow quickly from cut stumps.

Old trees are often a great size. One at Nikko, which Elwes photographed at the turn of the century, was 45m (150ft) tall and 6.9m (23ft) in girth at an estimated age of 300 years. Another at Koyasan measured 55m × 7.3m (180ft × 24ft). However, 33 to 36m (110 to 120ft) seems to be about the average height, with girths of 3 to 4.5m (10 to 15ft).

This tree is certainly long-lived. In an area of primeval forest on Yaku Island, south of Kyushu, the enormous old trees are thought to be over 1,000 years old.

Timber

The timber, which is reddish in colour, fragrant, durable, light, straight-grained and easily worked, is used for almost every purpose in Japan but especially for building and for making wine barrels. In China it is a similarly important timber. In remote areas the bark is used for roofing.

Introduction, performance and use as an exotic

The Japanese Cedar was discovered by J. Cunningham on Chusan Island, off the coast of Chekiang in 1701, but was not introduced into Britain until 1842 when Sir Everard Home sent seeds to Kew. The seedlings from these were kept in a greenhouse until 1847 and then planted out. One had grown to 7.8m (26ft) in height and over 60cm (2ft) in girth by 1880 but died later; none of these originals has survived. A later introduction was made by Fortune in 1844 by seed sent from Chekiang.

The first introduction from Japan was by Maries in 1879, so all the old trees planted before that date are almost certainly of Chinese origin. The finest of these which Elwes found and photographed was at Hempstead, Kent, planted before 1850. It measured 24.4m × 2.4m (80ft × 8ft) in 1904, and Mitchell found it to be 33.8m × 3.2m (111ft × 10ft 8in) in 1970. Of the 10 or so trees of Chinese origin which Elwes measured, all had been planted between 1847 and 1856, and ranged in height from 17 to 24m (56 to 80ft) by 1906. By 1970, when Mitchell measured them, they had grown to between 22 and 34m (72 and 111ft).

The tallest in the country in 1970 was a specimen at Woodhouse, Devon, which measured 37m × 2.7m (121ft × 8ft 11in). Another at Endsleigh, Devon, measured 36.6m × 4.2m (120ft × 14ft 2in) in the same year. In fact, the most vigorous growing and best trees are usually found in the west and south of England. Here most large gardens and collections have specimens of what is a very fine, quick-growing ornamental.

Some recent measurements include the following trees:-

Bodnant, Gwynedd, 28.6m × 3.3m (94ft × 11ft) 1981
Endsleigh, Devon, 39.6m × 3m (130ft × 10ft 4in) 1987
Fota, Co. Cork, 35m × 4.5m (115ft × 15ft 3in) 1987

Powerscourt, Co. Dublin, 26.8m × 3.8m (88ft × 12ft 10in) 1987
Windsor Great Park, 30m × 3.9m (100ft × 13ft 4in) 1979

The fine specimen in the Royal Forestry Society's Arboretum at Leighton, Welshpool, is around 30m (100ft) tall and girths more than 3.6m (12ft).

A number of cultivars have been introduced from Japan. The most often planted is 'Lobbii' which came here in 1853, before the type tree arrived. Its leaves are tufts at the ends of the branches. Two originals are to be found at Dropmore, Bucks., measuring 26m × 2.2m (85ft × 7ft 6in) and 20.7m × 2m (68ft × 7ft). Another cultivar is 'Elegans' which has permanent feathery juvenile foliage but can be up to 21m (70ft) tall. There are also some dwarfs and semi-dwarfs.

This tree sets seed in Britain.

It is extensively planted as a forest tree in many parts of the temperate world, as well as in the mountains in some tropical countries. In India, where it was introduced in 1844, it was planted over a large area in the hills around Darjeeling and Shillong at altitudes between 1,220 and 2,290m (4,000 and 7,500ft). It has been used for shelter-belts in New Zealand and as an ornamental as well as a forest tree, having a greater resistance to a fungal infection than the Monterey Cypress. It is also particularly well-suited for under-planting some of the indigenous trees at altitudes of 760m (2,500ft). In South Africa it has proved itself as a very successful forest tree at altitudes between 920 and 1,520m (3,000 and 5,000ft).

The Japanese Cedar was introduced into the USA in 1861 and in the east it is hardy as far north as Boston. The cultivars 'Elegans' and 'Compacta' are the main varieties planted, the latter originating in Paris in 1877.

Japanese Umbrella Pine

Sciadopitys verticillata

In its native Japan, the Umbrella Pine is a tall tree up to 36.5m (120ft) in height, with a substantial bole clad in reddish-brown bark which scales off in long, thin shreds. It supports short, slender branches. These bear the conic crown composed of leaves in whorls like the spines of an umbrella. Each leaf is glossy green, 12cm (5in) long and needle-like, grooved down the middle, being two leaves fused into one. (See illustration facing p.65.)

Distribution and habitat

The Umbrella Pine is exclusively native to Japan and is distributed sporadically in the mountains of central and southern Honshu at elevations of between 180 and 1,520m (600 and 5,000ft), mainly on rocky slopes and ridges. It is rarely found in pure stands but usually in mixture with Momi Fir *Abies firma*, Hinoki Cypress *Chamaecyparis obtusa*, South Japanese Hemlock *Tsuga sieboldii* and Hiba *Thuyopsis dolobrata*.

Life history

The 7cm (3in) long cones, with thick woody scales, ripen in October of the second year to shed their narrow, laterally winged seeds which germinate readily in dense shade and grow regularly but very slowly – 7 to 15cm (3 to 6in) annually for the first 20 years. In more open conditions they may attain 9m (30ft) in 50 years but trees of 30 to 36m (100 to 120ft) in height, which is about the maximum, will probably be at least 250 years old. Even in old age these trees retain their regular conic shape and rarely become flat-topped. In Japan the Umbrella Pine is widely planted in parks, gardens and temple grounds.

Timber

The wood of the Umbrella Pine is soft, white, elastic and very durable. It is used in Japan for bath tubs, casks and boat-building.

Introduction, performance and use as an exotic

The first specimen of the Umbrella Pine to reach Britain was a plant sent by Thomas Lobb from the Botanic Gardens at Buitenzorg, Java, to Veitch's Exeter nursery in 1853. It did not survive for long. The main introduction was by seed sent from Japan by J.G. Veitch in 1861. In 1876, some of the plants from this seed started to bear cones when only 90cm (3ft) tall; and ripe seeds were produced by a tree at Castlewellan, Co. Down, in 1900 but seedlings raised from these by Elwes were always weak and died after two or three years. Plants raised from seed sent to him from Japan fared little better.

Elwes records only one specimen, a tree at Benenden School, Kent. This he photographed in 1905 when it was 11.5m × 0.6m (38ft × 2ft). It is still there and is by far the tallest Umbrella Pine in Britain today. Mitchell measured it in 1970 as being 22.5m × 1.3m (74ft × 4ft 8in), and again in 1987 when it had grown to 22.8m × 1.5m (75ft × 5ft). Its planting date is not known. Another tree measured by Mitchell in 1970 was at Castle Kennedy, Wigtownshire. This was planted in 1878 and measured a mere 12.2m × 0.8m (40ft × 2ft 9in) in 1967.

Trees measured more recently include:-

Armadale Castle, Skye, 12.2m × 1.8m (40ft × 6ft 3in) 1987
Bodnant, Gwynedd, 20m × 1.5m (66ft × 5ft 1in) 1987
Killerton, Devon, 14m × 1.5m (46ft × 5ft 3in) 1980
Mt Usher, Co. Wicklow, 21.3m × 1.1m (70ft × 3ft 10in) 1987
Sheffield Park, Sussex, 17m × 1.6m (56ft × 5ft 6in) 1988
Tilgate, Surrey, 15.2m × 1.5m (50ft × 5ft) 1988

Although generally slow growing compared with many other conifers, the Umbrella Pine is not always so. Mitchell records height growth of 30cm (1ft) a year over long periods with attendant increase in girth of 2.5cm (1in) a year.

The tree is not common in Britain but most large collections have some specimens, always neat, well furnished and immaculately regular in growth.

In North America it is a popular ornamental due to its unusual leaf formation and its regular shape. Again, in that country, it tends to be a slow grower and seems to do best along the Atlantic Coast, particularly on Long Island and also in the Pacific coastal area. It can stand temperatures down to -23C (-10F); lower temperatures tend to brown the foliage.

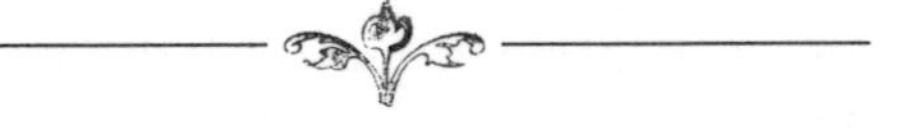

Chinese Fir

Cunninghamia lanceolata

The Chinese Fir is a large, evergreen tree, up to 45m (150ft) tall, exclusively native to China, with a domed crown of short, drooping branches. These bear flexible, prickly, lance-shaped needles densely arranged on the branchlets and generally spreading laterally either side, somewhat untidily arranged. The straight bole has stringy, reddish-brown bark, marked with narrow, vertical fissures. (See illustration facing p.65.)

Distribution and habitat

The natural range of the Chinese Fir is central and southern China south of a west-east line from Szechuan, Hupeh and Honan to Kwangsi, Kwangtung, Fukien and Hong Kong. Here it is found, often in pure stands mainly in the mountain valleys enduring hot summers and high humidity. It also grows down at sea level in the south and as high up as 1,520m (5,000ft) in the central area.

It is a component of the mixed temperate forests which range over extensive areas, and occurs especially in places where interference by man has been severe with secondary forest gradually re-appearing. Its associates include South China Pine *Pinus massoniana*, Weeping Cypress *Cupressus funebris*, Armand's Pine *Pinus armandii*, Tulip Tree *Liriodendron chinensis* and Chinese Hemlock *Tsuga chinensis*, at lower levels; and Japanese Cedar *Cryptomeria japonica*, Sweet Gum *Liquidambar formosana* and Camphor Tree *Cinnamomum camphora* at lower levels in the south.

After a visit to Hong Kong in 1977, I wrote in *Quarterly Journal of Forestry*:-

> 'Indigenous woodland is scarce (in Hong Kong). What little there was fell to firewood hunters during the Japanese occupation, and later during the great influx of refugees in the early fifties. But there is one officially designated 'Special Area' which, since the war, has been protected from any kind of exploitation or intrusion. Here something approaching the indigenous forest cover can be seen. The hillsides are now clad in a dense and luxurious mixture of which I could not begin to identify. However, I was delighted to see, growing in considerable profusion, little groups of Chinese Firs *Cunninghamia lanceolata*, some possibly natural seedlings, others more likely coppice shoots from old stumps, but natural none the less . . . Often in England cunninghamias look rather scruffy and weary of a climate not at all to their liking; but growing amongst a host of co-patriots they looked, as they were, quite at home and one could quite understand why they are one of China's most important timber trees.'

This tree is also extensively planted in Chinese re-afforestation schemes, as it readily produces coppice shoots when cut.

Life history

The squat, round female cones are borne at the ends of the branches, and three seeds are produced at the base of each thin, woody scale.

The seedlings, which are very frost tender, grow 7 to 10cm (3 to 4in) in the first year and the growth rate increases especially after they have attained the first 90 to 120cm (3 to 4ft). In maturity some trees may be as much as 45m (150ft) tall with girths of up to 5.4m (18ft). In forest stands the superbly straight trunks are often free of branches for 12m (40ft) or more.

Timber

The reddish-brown, fragrant, fine-grained and durable timber is in great demand in China for house building, general carpentry, tea chests and many other minor uses. In some mountain areas the bark is used for roofing. Old logs, buried by earthquakes and landslides for 200 years are, when dug up, found to be in an excellent state of preservation and are greatly valued in China for top quality coffins.

Introduction, performance and use as an exotic

The Chinese Fir was discovered by J. Cunningham on Chusan Island in 1701, and his first specimen is in the British Museum. Lambert was the first to describe the tree from other specimens sent home in 1793 by Sir George Staunton. It was finally introduced into Britain from Canton by William Kerr in 1804, but most of the early trees now growing in the country were in fact from seed collected by Fortune in 1844.

The trees from Kerr's seeds were initially grown in greenhouses, and one of the first to be taken outside was at Claremont, planted in 1816. It was 5.4m (18ft) tall in 1837 when Loudon first saw it. The same tree was measured by Mitchell in 1965 when it was 15.8m × 1.8m (52ft × 6ft 2in). Two other specimens from the original seed were planted outside at Dropmore, Bucks., in 1822, and had grown to some 5m (17ft) in height by 1837. In 1970 they were 16.4m × 1.4m (54ft × 4ft 10in) and 15.8m × 1.2m (52ft × 4ft 4in).

The finest which Elwes found was at Killerton, Devon; it measured 19m × 1.2m (62ft × 4ft) in 1904 and by 1977 it had grown to 26m × 2.4m (85ft × 8ft). The tree at Bicton, Devon which Elwes found to be 17m × 1.4m (56ft × 4ft 10in) in 1906, was, according to Mitchell, the tallest in Britain in 1970 when it measured 32m × 2.3m (105ft × 7ft 10in). But in 1977, after damage, it was only 26m × 2.4m (85ft × 8ft). Another tree at Pencarrow, Cornwall, planted in 1850, and measured by Elwes in 1905 as 12.2m × 1.3m (40ft × 4ft 8in), had grown to 24m × 2.6m (79ft × 8ft 8in) by 1970, and then had the largest girth of the species in the country. By 1975 it had grown a little to 24.4m × 2.6m (80ft × 8ft 10in).

Of the 24 trees measured by Mitchell in or before 1970, including the above, most seem to have grown about 22cm (9in) a year over a century or so, and nearly all of them are in the south of England.

Other trees measured fairly recently include:-

Holmbush, Sussex, 15.8m × 2.4m (52ft × 8ft 1in) 1988
Mt Usher, Co. Wicklow, 26.8m × 2.8m (88ft × 9ft 7in) 1987
Oakly Park, Salop, 21.3 × 1.9m (70ft × 6ft 6in) 1978
Tregrehan, Cornwall, 21.3 × 2.6m (70ft × 8ft 9in) 1987

The Chinese Fir is an interesting and, in early years, a handsome and attractive ornamental tree, although a tendency to suffer in spring frosts can make it look untidy. It should not be planted north of the South Midlands.

Trees from Hong Kong have been grown successfully in Malaya, and trial plots have been established in South Africa at altitudes of between 1,220 and 1,520m (4,000 and 5,000ft), but generally speaking it is cultivated as a forest tree only in its native China.

It is grown as an ornamental in Vancouver, British Columbia; and in the USA, where it was introduced in 1804, it is planted as an ornamental in the warmer parts of the country. Its ability to sprout from cut stumps is a point in its favour with gardeners.

Swamp or Bald Cypress

Taxodium disticum

The Swamp Cypress is one of the most unusual trees found in the south of North America. A deciduous conifer, at home on land or in water, it has a massive, buttressed, tapering trunk. This is surrounded by curious cone-shaped knots or 'knees' which arise from the stout, wide-spreading roots. Old trees are often found growing to heights of 36m (120ft) or more. The trunk supports a broad irregular crown which can be 30m (100ft) across with a light-green, feathery foliage which turns foxy brown in the autumn before falling. (See illustration facing p.80.)

Distribution and habitat

The natural range of the Swamp Cypress extends along the coastal plain from Delaware in the north, south to Florida and west as far as the Mexican border, as well as inland along many streams of the south eastern States and up the Mississippi Valley as far as south east Oklahoma, south east Missouri, south Illinois and south west Indiana.

The tree is usually restricted to very wet soils where moisture is plentiful and permanent. Such sites are found mainly on flat ground at elevations of less than 30m (100ft), but in the Mississippi Valley the tree is occasionally found as high up as 152m (500ft). Although the best growth seems to occur on well-drained but moist soils, it is rarely found there because of competition from other species on such sites.

On the damper flood plain sites the Swamp Cypress associates with such trees as Water Tupelo *Nyssa Aquatica*, Black Willow *Salix nigra*, Swamp Cottonwood *Populus heterophylla*, Water Hickory *Carya aquatica* and Pond Pine *Pinus serotina*; and in the drier slightly elevated areas in swamps with Slash Pine *Pinus elliotii*, Laurel Oak *Quercus laurifolia*, and Sweet Gum *Liquidambar styraciflua*.

Life history

The spherical cones, green at first, become brown as they ripen in the autumn. Their 30 or so seeds are dispersed by flood water washing them down stream. The seeds will germinate only in an abundant supply of water but not if submerged, although they can remain under water for up to three years, germinating when the floods recede. Growth is rapid, for seedlings must be tall enough at the end of the first year to stay above the floods – usually 20 to 50cm (8 to 20in) – and double that by the end of the second season.

Later growth rates compare favourably with competing species – about 30cm (1ft) a year for the first 100 years. These trees live to a great age, in virgin stands varying between 400 and 600 years, with individuals as old as 1,200 years, particularly in Georgia and South Carolina. One specimen in Tennessee was reported to be 1,300 years old, 37m (122ft) tall and nearly 12m (40ft) in girth.

Although seedlings can develop in heavy shade, the best stands start in clearings.

Timber

The timber is pale brown to reddish in colour and is easily worked, with no resin ducts. It is very durable and is used for flooring, water tanks, coffins, shingles and boats. It has a somewhat rancid smell.

Introduction, performance and use as an exotic

The Swamp Cypress was probably introduced into Britain before 1640, as Parkinson, writing in that year says '. . . its seed was brought here by Master Tradescant from Virginia and sown here and does spring very bravely'.

Loudon recorded a number of trees growing in England in 1838, the tallest being at Whitton, near Hounslow, which topped 24.7m (81ft). This tree was probably one of the group which Elwes photographed in 1905 the tallest of which was 29.8m × 3.7m (98ft × 12ft 6in). Another specimen which Loudon mentions, at Stratsfield Saye, measuring 14m × 1m (46ft × 3ft 4in) in 1838, Elwes found to be 19.2m × 2.7m (63ft × 9ft) in 1903; and Mitchell measured it in 1968 as 25m × 3.8m (82ft × 12ft 10in).

The tree at Longford Castle, Wilts., which Elwes found to be 'very tall' and girthing 2.8m (9ft 10in) in 1904 was, according to Mitchell, the tallest in Britain, at 34m × 3.6m (112ft × 12ft) in 1962, but seems to have been surpassed by one at Broadlands, Hants., which, in 1968, was 35.6m × 4.3m (117ft × 14ft 9in), and in 1976, 36m × 4.5m (118ft × 15ft 5in).

Two trees at Dean Court, Wimborne, Dorset, and at Burwood Park, Surrey, which measured 30m × 5.3m (100ft × 17ft 8in) and 32m × 5.2m (105ft × 17ft 7in) respectively in 1968 and 1965, are so much greater in girth than any others known in Britain that Mitchell is tempted to wonder if they are from Tradescant's original seed, for they are remarkably similar in shape and size. In 1973 the Burwood tree had grown to 35m × 5.4m (115ft × 18ft 3in).

More recent measurements include the following trees:-

Fawley Court, Oxon, 26.8m × 4m (88ft × 13ft 6in) 1987
Forde Abbey, Dorset, 29m × 3.2m (95ft × 10ft 6in) 1987
Kenwood, London, 22.8m × 3.7m (75ft × 12ft 6in) 1987
Knaphill, Surrey, 32m × 3.6m (105ft × 12ft) 1987
Pusey House, Oxford, 29m × 4.5m (95ft × 15ft) 1988
Rectory, Much Hadham, Herts., 31m × 5.2m (102ft × 17ft) 1987

The Swamp Cypress is a very healthy tree in Britain and even the oldest specimens, many of which exceed 200 years of age, show no signs of debility or disease. They seem to grow best when planted beside ponds and lakes but will thrive in drier situations. Although they can stand low temperatures in the winter, they need warmth for proper growth and are thus found mainly in southern England, especially round London. They make splendid ornamental trees, with their fine shape, delicate greens in the spring and summer, and their beautiful russet hues in the autumn.

In the USA, this tree, native to the south-eastern areas, is hardy as a planted ornamental as far north as Boston where, according to Wyman, there is a fine specimen 24m (80ft) tall. A cultivar 'Pendens', with pendulous branches, is also planted.

Dawn Redwood

Metasequoia glyptostroboides

The Dawn Redwood is a beautiful deciduous Chinese conifer, discovered as recently as 1941 and before that known solely from fossil records. The bole, which is often fluted at the base, is covered in orange-brown, stringy bark. The trunk supports rather sparsely spaced, upswept branches which bear soft, bright green comb-like leaves. These are set opposite one another, unlike those of the similar Swamp Cypress, which are set alternately. The leaves turn various shades of pink and russet in the autumn. (See illustration facing p.80.)

Distribution and habitat

It is remarkable that in a country like China, where civilisation has so long prospered and large numbers of plant-hunters made such extensive excursions through the 19th century, any tree could have escaped detection. But it was not until 1941 that anyone knew of or even suspected the existence of the Dawn Redwood. At that time the Chinese Government was hard pressed by the Japanese and had retreated into the mountains in the west of the country, dedicating itself to a prolonged resistance. Exploring parties were sent out to survey and search little-known regions for natural resources. During one of these forays a botanist, called T. Kan, of Central University, noticed a peculiar deciduous conifer in East Szechuan which the locals called 'Water Fir'. No specimens were then collected, but in 1944 T. Wang, of the Forest Research Bureau, collected foliage and cones in the same area. Thus the discovery was made known to the whole world.

The Dawn Redwood is found growing naturally in a relatively small area of some 480 square kilometres (300 square miles) in west Central China; in the north west corner of Hupeh Province, and spreading at a few points slightly across the Szechuan border. The centre of this distribution, where it grows most strongly, is the Shui-Sha Valley – a narrow, close depression varying in altitude between 660 and 1,300m (2,150 and 4,250ft). Here it grows in shady, moist localities, mainly in ravines, both high up and on the valley bottom; but it thrives best along the rocky banks of small streams, and where seepage collects at the base of steep slopes. In these situations it forms dense thickets with such trees as Chinese Fir *Cunninghamia lanceolata*, Cow's Tail Pine *Cephalotaxus fortuni*, Sweet Gum *Liquidambar formosana* and many others.

Within this area of natural distribution there has been considerable planting of the Dawn Redwood; mostly the transplanting of natural saplings along the edges of rice fields, along roadsides and around homesteads, always for amenity, as the timber seems to be little valued.

Introduction, performance and use as an exotic

Between 1946 and 1948 several expeditions went into China to collect seeds and herbarium specimens. Through the efforts of Dr E.D. Merrill, of the Arnold Arboretum, Washington, D.C., a large quantity of seed was obtained in 1948 and distributed widely in the USA. Some was sent to Britain and was distributed to botanic gardens and horticultural societies.

Early growth was immediately found to be rapid. A plant from the original seed at the Morris Arboretum in America was 1.7m (5ft 9in) tall at the end of its second grown season, in 1949; by 1957 it was over 6m (20ft) tall. Other specimens from the original seed sown in America, put on as much as 1.3m (4ft 6in) in a single season.

Mitchell lists 28 original trees in Britain, all grown from the 1948 seed, sown in that year or in 1949. Nearly all of those growing in a damp woodland or standing near water had reached a top height of between 14 and 17.3m (46 and 57ft) by 1970; the tallest being a tree in the Savill Gardens, Windsor, which measured 17.3m × 0.9m (57ft × 3ft 1in) and the largest in the Cambridge Botanic Gardens measuring 15.5m × 1.3m (51ft × 4ft 5in). By 1981 the Savill Gardens tree had grown to 24.4m × 1.5m (80ft × 5ft) and the Cambridge tree, in 1987, was 22m × 2.6m (72ft × 8ft 9in). Trees planted on drier sites, although they grew rapidly to start with, have in general not done so well.

Recent measurements show how strongly this tree is growing:-

Bodnant, Gwynedd, 21.3m × 2.4m (70ft × 8ft 2in) 1987
City Park, Bournemouth, 26.8m × 3.2m (88ft × 10ft 9in) 1987
Clare College, Camb., 18.3m × 2.7m (60ft × 9ft 3in) 1987
Leonardslee, Surrey, 31m × 1.8m (102ft × 6ft) 1988
Oxford Botanic Garden, 20m × 2.4m (66ft × 8ft 3in) 1987
Penrhyn Castle, Gwynedd, 21.3m × 2.5m (70ft × 8ft 7in) 1987
Snowdenham House, Surrey, 22.8m × 2.4m (75ft × 8ft 2in) 1987
Wisley, Surrey, 26m × 1.9m (85ft × 6ft 7in) 1988

Although the Dawn Redwood has only produced an occasional female cone and never any male flowers in Britain, and thus no seed (except in controlled conditions under stimulation by gibberellic acid), the ease with which it can be propagated by cuttings (taken in June or July) has meant that the tree has already become widespread in Britain. The speed with which these cuttings grow has made its success that much greater. For instance, according to Mitchell, a rooted cutting planted beside a stream near Guildford, Surrey, reached 15.2m × 1.2m (50ft × 4ft 2in) in 18 years.

The beautiful green foliage which turns brick red in the late autumn, after many other trees have finished their display, the perfect narrowly conic shape and the orange brown bark and twigs in the winter, all combine to make this a most attractive ornamental tree. Add to this the fascination of its recent discovery and speculation about its future size, timber quality and general performance, and it becomes one of the most intriguing trees in the world.

In the USA where the first-ever seeds to come to the West were received at the Arnold Arboretum in 1948, the Dawn Redwood has also shown itself to be a very quick grower in moist soils. A tree at Wilmington, Delaware, grew 15m (50ft) tall from seed in 15 years. But its tendency to continue in growth into early autumn has made it vulnerable to early frosts in colder areas. Nevertheless, this fast growth, attractive foliage and its fascinating history have made it a popular ornamental in North America.

Veitch's Silver Fir

Abies veitchii

Veitch's Silver Fir is a tree of the mountains of central Japan and one of the smallest members of the genus in that country, rarely exceeding 23m (75ft) in height. A greyish-white bole, never very massive and often fluted, supports a conic, symmetrical crown of dark glossy green needles, all pointing forward and bending upwards to show the vividly white bands on the undersides, giving the whole tree a frosty appearance.

Distribution and habitat

The natural range of Veitch's Silver Fir is restricted to the sub-alpine region of Central Honshu in Japan, with one or two small outlying areas in Shikoku where a variety with shorter, stiffer needles grows in small numbers. Throughout its range this fir is rarely found below 1,520m (5,000ft) and it attains its finest proportions between 1,830 and 2,440m (6,000 and 8,000ft) on the south and south-west slopes.

Swamp Cypress, Corkscrew Reserve, Florida (p.76)

Dawn Redwood, Botanic Gardens, Oxford, planted 1949 (p.78)

Grand Fir, Yattenden, Hants. (p.82)

Colorado White Fir, Siskiyou Mts. Calif.
(p.85)

Noble Fir, Cragside, Northumberland
(p.87)

Korean Fir, Queenswood, Herefords.
(p.89)

Red Fir, Cragside, Northumberland
(p.90)

This fir is sometimes found in pure stands but is more often intermingled with other trees such as Hondo Spruce *Picea hondoensis*, Maries Fir *Abies mariesii* and Northern Japanese Hemlock *Tsuga diversifolia*.

Life history

The flowers, which are borne early in life before the tree is 6m (20ft) tall, open in June ahead of the leaf flush. The cylindrical female cones gradually grow to about 7cm (3in) tall and turn a dark purple when ripe in October, releasing broad-winged seeds.

Growth is very rapid in the early stages – some 3 to 3.6m (10 to 12ft) in 5 years – and even more so during the next 15 years when leading shoots 90cm (3ft) long are common; but the tree is short lived and quickly loses its symmetrical habit.

In maturity, top heights of 23m (75ft) and girths of up to 2.5m (8ft) are usual, with a very occasional specimen reaching 30m (100ft).

Timber

The white timber is of commercial value only in Japan where, in the trade, it is not distinguished from that of the Momi Fir *Abies firma* and is put to the same uses – constructional work, coffins, boxes and pulp.

Introduction, performance and use as an exotic

Veitch's Silver Fir was discovered by J. Gould Veitch, the Exeter nurseryman, on Mount Fujiama in 1860, but seems to have been introduced into the USA before Britain, when a Mr. Hogg sent some specimens to Parson's Nursery, New York in 1876. One of these, according to Elwes, had grown to a height of 4.5m (15ft) by 1879, the same year in which Maries, Veitch's collector, sent seeds to England. The ensuing seedlings came through the very severe winter of 1879-80 unscathed.

Elwes knew of no specimen of this fir of any size. One at Tregrehan, Cornwall, was 8.2m × 0.6m (27ft × 2ft) in 1908 and another in Ochtertyre, Perthshire, which was 9m × 0.6m (30ft × 2ft) in the same year. Bean mentions one at Murthly Castle, Perthshire, which was 9.4m (31ft) tall in 1906.

It seems that none of these early trees has survived to this day, but Mitchell reports that an original tree from Veitch at Eridge Castle, Kent, planted in 1881, died in 1951, when it measured 20.7m × 2.2m (68ft × 7ft 8in). The most vigorous and healthy tree Mitchell had seen was one at Avondale, Eire, planted around 1909, which was 22m × 1.1m (72ft × 3ft 10in) in 1968. There is one at Westonbirt, Glos., planted in 1916, which is taller – 24m (79ft) – and there are two older ones, planted in 1890, at Borde Hill, Sussex, which were 21m × 1.4m (69ft × 4ft 8in) and 21.3m × 0.9m (70ft × 3ft 1in) in 1968. In the last decade or so some of the larger specimens up and down the country have made significant growth as more recent measurements show:-

Alnwick Castle, Northumberland, 28m × 1.8m (92ft × 6ft) 1977
Crarae, Argyll, 20m × 2.1m (66ft × 7ft 4in) 1987
Glentanar, Aberdeen, 21m × 2.2m (69ft × 7ft 6in) 1987
Hergest Croft, Herefords., 25m × 2.2m (82ft × 7ft 5in) 1981
Ochtertye, Perths., 26m × 2.4m (85ft × 8ft 3in) 1987
Scone Palace, Perth, 22.2m × 2.1m (73ft × 7ft) 1987

This silver fir is an attractive ornamental which grows quickly to form a symmetrical tree with silvery-green foliage and beautiful purple cones which it produces freely in Britain, the seeds being viable. It is quite often grown in small gardens as well as in parks and collections, but is short-lived and tends to become thin and lanky in middle and old age.

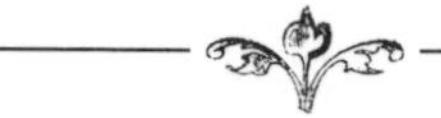

Grand Fir

Abies grandis

One of the giants of the north west American conifer forests, the Grand Fir, frequently exceeds 60m (200ft) in height and is sometimes as much as 75m (250ft). It stands alone or in open forest, its downward sweeping branches, with upturned ends, reaching almost to the ground in a gently tapering habit. The needles, silvery white below, are up to 5cm (2in) long and are flatter, more comb-like and glossier than those of other American firs. In the denser forest, the Grand Firs are usually dotted about among the other conifers. The great trunks, clad in thick reddish-brown bark, cut by narrow fissures into sharp horny ridges and oblong plates, are often free of branches for 30m (100ft) or more, the gradual taper emphasising their massive bulks. The bark of young trees is smooth and ashy-brown in colour, with chalky marks and resin blisters here and there. (See illustration facing p.80.)

Distribution and habitat

The natural range of the Grand Fir is divided into two distinct zones: in the Pacific coastal region, from the east side of Vancouver Island and the adjacent mainland of British Columbia, south through western Washington and Oregon, to northwest California; and in the continental interior, from the Okanagan and Kootenay Lakes in southern British Columbia south to eastern Washington, central Idaho, western Montana and northern Oregon. The largest trees are found in western Washington and on Vancouver Island, and in the Redwood region of California, growing in deep, rich alluvial soils along streams and valley bottoms at elevations between sea level and 300m (1,000ft). In these coastal regions, it is predominantly a lowland species, but in the interior where some of the best commercial stands are to be found, particularly in northern Idaho, the trees grow well at between 1,220 and 1,520m (4,000 and 5,000ft).

On the Pacific Coast, the Grand Fir grows in association with Sitka Spruce *Picea sitchensis*, Pacific Silver Fir *Abies amabilis*, Western Hemlock *Tsuga heterophylla*, Red Cedar *Thuya plicata*, Douglas Fir *Pseudotsuga menziesii*, Red Alder *Alnus rubra* and Black Cottonwood *Populus trichocarpa*; and in the interior with Sugar Pine *Pinus lambertiana*, Incense Cedar *Calocedrus decurrens* and Oregon White Oak *Quercus garryana*.

Life history

Grand Firs start producing their small, upright green cones at the age of about 20 years and production increases with age up to 100 years and more. Good seed years – rated to be when each tree on average produces more than 40 cones – are rather sporadic and only about half as frequent as with Western Hemlock and other associated trees, with little more than a quarter of the seed being viable. Of the consequent seedlings, more than a third die in the first season. However, in moderate shade, if free from excessive heat and drought, the surviving seedlings are aggressive enough to form a considerable proportion of the mixed seedling stands in which they are growing. Early height growth equals that of Douglas Fir and, on Vancouver Island, terminal shoots 90cm (3ft) long and trees 42.5m (140ft) tall at 50 years of age have been recorded. In the interior though, annual height growth may be as little as 12cm (5in).

On optimum sites in the coastal region, Grand Firs are usually 45 to 60m (150 to 200ft) tall with girths of 3m (10ft). The largest tree recorded, on Vancouver Island, was 73m (240ft) tall, girthed 4.5m (15ft) and was 280 years old, which suggests that these firs are not very long lived when compared with Redwoods and Douglas Firs.

Grand Firs are rated as shade tolerant, more so than Douglas Firs, and can survive for many years suppressed in dense shade, quickly building a crown when released.

Timber

The timber of Grand Fir is soft, straight grained and yellowish brown in colour; but because of its brittleness, lack of strength and durability it has few uses other than for pulp and such minor things as packing cases, crates and rough construction timber.

Introduction, performance and use as an exotic

The Grand Fir was discovered by David Douglas in 1825, but seed was not sent to Britain until 1831 and from this very few plants were raised, although, according to Murray, 'a multitude of young plants existed which had been raised from cuttings' (no easy thing to do with *Abies* spp.). However, a second introduction took place when William Lobb sent seed to Veitch in 1851, and Jeffrey did likewise to the Scottish Oregon Association in the same year. Mitchell records that one original Douglas tree exists at Curraghmore, Co. Waterford, which was planted there in 1835 and was 32m × 5.2m (105ft × 17ft 5in) in 1968. He reckons that another tree at Lochanhead, Dumfries, which measured 33.5m × 6m (110ft × 19ft 9in) in 1962, could be another Douglas specimen as it has a greater girth than any other Grand Fir in Britain, bar one, and is not on a particularly favourable site.

Most of the biggest specimens, and there are many over 45m (150ft) tall and 4.5m (15ft) in girth, were planted after 1870. In 1969 the tallest in Britain was a tree at Strone, Argyll, planted in 1876, which topped 53m (175ft) and girthed 4.9m (16ft 8in), having put on 22.5m (74ft) in height and 1.5m (5ft) in girth since being measured in 1931. By 1976 it had grown to 57.3m × 5.3m (188ft × 17ft 6in). By 1987 it had become one of the few trees in Britain to exceed 60m (200ft) in height measuring 62.4m × 5.7m (208ft × 19ft).

Other large trees recently measured include:-

Balmacaan, Inverness, 51.8m × 7m (170ft × 23ft) 1988
Blair Castle, Perth, 54m × 5.4m (177ft × 18ft), 55m × 4.8m (180ft × 16ft) 1987
Charleville, Co. Wicklow, 39.6m × 5m (130ft × 16ft 7in) 1987
Fairburn, Ross & Crom., 51.8m × 4.9m (170ft × 16ft 6in) 1987
Fota, Co. Cork, 42m × 5.6m (138ft × 18ft 10in) 1987
Murthly Castle, Perth, 41m × 6.8m (135ft × 22ft 8in) 1987

As a forest tree in Britain the Grand Fir is steadily increasing in popularity. It is healthy over a wide range of rainfall, doing well with only 760mm (30in) but best where precipitation exceeds 1,000mm (40in) annually, although it is never so demanding of moisture as Sitka Spruce. It seems to be equally productive in the cold north as it is in the warmer south, but it is very sensitive to exposure and ill-adapted to peaty or leached sandy soils, and those of a highly calcareous character.

Early growth in Britain is rapid – 15m (50ft) in 20 years on good sites; a tree at Dunster, Somerset, reached 32m (106ft) in just 41 years. There seems to be little doubt that Grand Firs planted in sheltered sites where the wind cannot blow out their tops, will reach dimensions comparable with those attained in their natural range. Already the girths of some of them are tending to exceed those of the largest trees in the Pacific Coastal region.

The Grand Fir, because of its size and susceptibility to damage by the wind, is not nowadays so often planted as an ornamental. There are, however, some very large and handsome ornamental specimens all over Britain in large gardens and parks, mostly resulting from the early planting.

White Fir

Abies concolor

The White Fir is one of the most economically important of the firs growing on the Pacific Slope and the Rocky Mountains in north west America. Its massive proportions are emphasised by the heavily foliaged, conic crown of blue-grey needles. These needles stand almost erect on the twigs, supported on short, thick branches from the trunk which is covered in ashy-grey, rather horny bark. Young trees are exceptionally symmetrical in form. (See illustration facing p.81.)

Distribution and habitat

The natural range of the White Fir is extensive in the Pacific Coast region and on the Rockies, from New Mexico and Colorado westward to Oregon and California; and locally in northern Mexico.

It attains its largest sizes on moist soils on east facing slopes and at the heads of streams in California and south Oregon where it is occasionally found as low down as 610m (2,000ft). Further south, along the western slopes of the Cascades, it is found up to 1,830m (6,000ft) and on the eastern slopes at 2,290m (7,500ft). In north east California it grows at altitudes of 2,440m (8,000ft), and in the interior as high up as 3,350m (11,000ft).

In the north Pacific Coastal area associated trees include Grand Fir *Abies grandis*, Douglas Fir *Pseudotsuga menziesii* and Lodgepole Pine *Pinus contorta*; in central regions it grows with Ponderosa Pine *Pinus ponderosa*, Incense Cedar *Calocedrus decurrens* and Sugar Pine *Pinus lambertiana*; and in the south with Red Fir *Abies magnifica*, Wellingtonia *Sequoiadendron giganteum* and Blue Spruce *Picea pungens*. In the high Rockies it consorts with, amongst others, Sub-alpine Fir *Abies lasiocarpa* and Engelmann's Spruce *Picea engelmannii*.

Life history

The 7cm (3in) tall, upright cones reach maturity, disintegrate and shed their seeds in late September and October, good seed crops being produced every five years on average. The seed germinates in the spring and growth is very slow for the first 30 years. In Central California top heights of 43 to 55m (140 to 180ft) and girths of 3.5 to 4.5m (12 to 15ft) are usual, with the occasional tree up to 60m (200ft) tall and 5.4m (18ft) in girth. The American Forestry Association's champion White Fir measures 57.6m × 7.9m (189ft × 25ft 5in).

The White Fir does not appear to be very long lived, 350 years being about the average for mature trees, but some of the largest specimens may be as much as 500 years old.

It is shade tolerant and can usually compete adequately with most of its competitors, especially in early years.

Timber

The white, straight-grained, fine textured timber is not distinguished by the trade from that of the other Western American firs. It is used for general construction work, boxes, crates and pulp.

Introduction, performance and use as an exotic

The White fir was originally introduced into Britain as two separate species – White Fir *Abies concolor* and Low's Fir *Abies lowiana* – but since then the latter has been distinguished only as a variety, *A. concolor* var. *lowiana*, although still called Low's Fir. The type, the White Fir, is the tree which grows at high altitudes in southern California and Colorado; and Low's Fir in northern California and Oregon at lower elevations. Low's Fir has itself been divided into two forms; one which more closely resembles Grand Fir, with flatter needles; and the other approximates to the White Fir in the upright setting of its needles.

The White Fir was introduced into Britain in 1873 when seeds and specimens were sent from New Mexico to a nursery at St. Albans, Herts. Elwes found very few White Firs of any size at the turn of the century, the one at Highnams Court, Glos., measuring 13.4m × 0.8m (44ft × 2ft 9in) in 1908, being the tallest. Large specimens do exist today, especially in Northumberland, Argyll and Perthshire. In the south and west they seem to lose their shape and in the south and east they do not grow well at all. The finest one which Mitchell has seen is at Cragside, Northumberland which, in 1958, was 40m × 3.5m (131ft × 11ft 6in); and, in 1974, 47m × 3.6m (154ft × 12ft 1in). A tree at Capenoch, Dumfries, planted as late as 1926, measured 26.8m × 2.2m (88ft × 7ft 3in) in 1979.

Low's Fir was introduced earlier than the White Fir, by William Lobb, who sent seeds to Veitch's Exeter nursery in 1851. It seems to have fared much better in Britain than its near relative and over a wider area of the country. The finest specimen which Elwes photographed and measured was at Linton Park, Kent. He found it to be 26m × 1.6m (85ft × 5ft 6in) in 1902; and when Maynard Greville saw it in 1956 it had grown to 36.6m × 4.8m (120ft × 16ft 3in); the largest in the country, but it was felled soon after. Another tree, planted in 1866 and measured by Elwes at Youngsbury, Herts. was, in 1907, 20.7m × 1.6m (68ft × 5ft 6in); by 1966 Mitchell found it to have grown to 33m × 2.3m (108ft × 7ft 8in). The tallest Low's Fir which Mitchell has found is at Durris House, Kincardine, which had been 22.5m × 2.3m (74ft × 7ft 6in) in 1904; 46.6m × 4.6m (153ft × 15ft 7in) in 1970; and 50.3m × 4.9m (165ft × 16ft 4in) in 1980. Some more recent measurements include:-

Blair Castle, Perth, 53m × 3.8m (174ft × 12ft 10in), 32m × 4.7m (105ft × 15ft 5in) 1987

Durris House, Kincards., 42m × 4.8m (138ft × 16ft) 1987
Faskally House, Perths., 39.6m × 3.6m (130ft × 12ft 2in) 1987

As a forest tree, Low's Fir seems to be able to produce a big volume of timber on soils where the traditionally big producer, Grand Fir, cannot; thus it may have a place in forestry for high production on poor soils. It is less tolerant of exposure than Noble Fir *Abies procera* and is unlikely to replace it at higher elevations.

In the USA, of all the native firs, this species seems to be the hardiest for planting in northern and eastern areas as far north as Boston. It withstands both heat and drought, as well as cold. A variety *violacea*, with bluish-white needles, is very popular for landscape planting, being softer looking than the Blue Spruce *Picea pungens* 'Glauca'.

Noble Fir

Abies procera

This tree is a giant of the mountains of north west America and is, as its name implies, an aristocrat among firs. Its somewhat bushy, rounded crown of short rigid branches, held aloft 60m (200ft) or more on a clean symmetrical stem, is very distinctive from a distance. Close inspection of the twigs reveals blue-green needles, in vivid contrast to the delicate pea green of the spring shoots which hug the stem at their bases and then curve upwards. The bark of old trees is rather thin, reddish-brown in colour, and broken by narrow grooves into broad, flat ridges. In young trees, which are conical in shape, the bark is smooth greenish grey. (See illustration facing p.81.)

Distribution and habitat

The natural range of the Noble Fir extends from the slopes of Mt Barker in northern Washington south along the west side of the Cascade Range, and in parts of northern Oregon, to the Siskiyou Mountains in northern California; mostly on gently sloping ridges, in valleys and on plateaus from 610 to 2,140m (2,000 to 7,000ft). Only occasionally does this fir descend to low levels. It grows on a wide range of soils, the controlling factor for success seeming to be a good water supply rather than high soil fertility.

At some parts of its range the Noble Fir is associated with Douglas Fir *Pseudotsuga menziesii*, Pacific Silver Fir *Abies amabilis*, Western Hemlock *Tsuga heterophylla* and Red Cedar *Thuya plicata*; and higher up in the mountains with Mountain Hemlock *Tsuga mertensiana*, Sub-alpine Fir *Abies lasiocarpa* and, in the Siskiyous, with Brewer Spruce *Picea brewerana*.

Life history

The Noble Fir starts producing cones at 20 years of age in the forest and much earlier in planted trees. They are up to 15cm (6in) long, and as much as 25cm (10in) long in planted specimens, pale purplish brown in colour, with long bracts protruding and bending over the scales. During a good season a single tree may produce as many as 500 cones (one tree is recorded as having borne over 800) although these trees are only occasionally so prolific.

The seed has a low germination capacity and remains viable for only one season. Initial growth of seedlings is slow, 2.5 to 5cm (1 to 2in) in the first season being most usual, and they often take three to four years to reach a height of 30cm (1ft). This slow growth continues until the trees are 10 to 15 years old, when it accelerates and the Noble Fir is able to compete with and even overtake its associates. The largest Noble Fir on record is 79m (260ft) tall and 7m (23ft) in girth. Trees of this sort of dimension are rated to be 600 to 700 years of age – old for a true fir. In 1903, Elwes counted 360 rings on a stump 3.6m (12ft) in girth. This fir, not being very shade tolerant, is rarely found as a co-dominant, and its rather scattered distribution at lower levels suggests that it is gradually being ousted by competitors, such as Western Hemlock and Silver Fir. It does, however, seem to hold its own with Douglas Fir on some sites.

Timber

The timber of Noble Fir is hard, strong, firm and very close-grained. In colour, it is light brown and marked with reddish brown streaks which add to its beauty. It works easily and takes a polish. It is used for flooring, interior finish, window frames, boats, boxes and crates.

Introduction, performance and use as an exotic

The Noble Fir was discovered by David Douglas in 1825 on the south side of the Columbia River, but it was not introduced by him to Britain until after his second journey in that area in 1830. None of the original trees now exist: the one at Dropmore, Bucks., measured by Mitchell in 1961, when it was 25.9m (85ft) tall and girthed 3.3m (11ft 2in), died in 1968. Later and more successful importations of seed came in from 1850 onwards and, by the time Elwes was writing at the turn of the century, there were many fine trees up and down the country between 24 and 30m (80 and 100ft) tall. A few of these remain and have all added 6 to 9m (20 to 30ft) to their height. Perhaps the tree at Blair Castle, Perth, planted by the Duke of Atholl in 1867, and measured by Elwes in 1904 as 21.3m × 1.5m (70ft × 5ft), is the same tree as the one Mitchell found there in 1981 to be 37m × 5.4m (121ft × 17ft 8in). A great many others, unrecorded by Elwes, presumably because they were too small at that time, are now well over 30m (100ft) tall. In 1970 the tallest, according to Mitchell, was a specimen at Inverary, planted in 1878, which topped 44m

(147ft) and was 3.3m (11ft 3in) in girth. Indeed, most of the large trees are in Scotland where they appear to suffer less from exposure than elsewhere. It seems unlikely that Noble Firs will ever much exceed 45m (150ft) in height anywhere in Britain because of their habit of coning heavily near the tops, followed by breakage from sheer weight in autumn gales.

Recent measurements include the following trees:-

Blair Castle, Perth, 36m × 5.7m (118ft × 18ft 9in), 27.4m × 5.4m (90ft × 17ft 10in) 1987
Durris House, Kincards., 45.4m × 5m (147ft × 16ft 9in) 1987
Fairburn, Ross & Crom., 36m × 5.4m (118ft × 18ft 3in) 1987
Glenferness, Nairn, 32m × 5.6m (105ft × 18ft 10in) 1987
Scone Palace, Perth, 37m × 4.7m (121ft × 15ft 6in) 1987

As a forest tree the Noble Fir is not as much planted as Grand Fir. It is, for one thing, less shade tolerant and thus less suitable for underplanting. On the other hand, its tolerance of exposure, particularly at high altitudes, makes it a successful species for planting on open moorland. In its early years it is a slow grower in Britain, just as it is in America, but even on exposed sites at heights of 300m (1,000ft) it can eventually average about 37cm (15in) a year for the first 30 years. On such sites, this compares very favourably with Sitka Spruce and the fir could well replace it on certain of the more exposed and drier sites.

Korean Fir

Abies koreana

The Korean Fir is an alpine species and the smallest of the firs, rarely exceeding 15m (50ft) in height and is often very much smaller. The conic crown is beautifully symmetrical and the level branches carry short, thick-set, blunt, dark green needles with whitish undersides, radiating haphazardly from the twigs. (See illustration facing p.81.)

Distribution and habitat

The natural range of the Korean Fir is restricted to South Korea and the volcanic island of Quelpeart some 64 kilometres (40 miles) south of the peninsula, where it forms considerable forests on the steep hillsides, usually in pure stands at altitudes above 920m (3,000ft).

Life history

The purplish-blue 5-7cm (2-3in) tall cylindrical cones are produced in exceptional profusion on even very small trees no more than 90cm (3ft) tall. They exude a protective resin and produce large, blue winged seeds which are distributed by wind.

Introduction, performance and use as an exotic

The Korean Fir was discovered by a Frenchman called Pierre Faurie on Quelpeart Island in 1908. It came to Britain via France in 1913 when the first specimen was planted in the Royal Botanic Gardens, Kew.

Although, in cultivation, it is usually seen as a semi-dwarf, it is by no means always so. In 1970 Mitchell listed ten specimens up and down the country which varied in heights between 8 and 11m (27 and 37ft) and girthed between 60 and 90cm (2 and 3ft); all were 30 to 40 years old.

A specimen at Crarae, Argyll, which he measured as 11.5m × 0.9m (38ft × 3ft) in 1976, had grown to 12.2m × 1m (40ft × 3ft 4in) ten years later. Other recent measurements include:-

Borde Hill, Sussex, 12.5m × 0.6m (41ft × 2ft 2in) 1981
Durris House, Kincards., 12.8m × 1.6m (42ft × 5ft 6in) 1987
Hergest Croft, Herefords., 11.5m × 1m (38ft × 3ft 6in) 1978

The Korean Fir is a most attractive and hardy subject for small gardens, not only because of its neat and compact habit, but also because of the precocious and profuse cone production which gives even small trees, no more than 90cm (3ft) tall, an added and interesting attraction each summer.

This fir was introduced into the USA in 1918. As a slow grower with a relatively small maximum height it is a popular subject there for small gardens.

Red Fir

Abies magnifica

This is a tree of high, mountainous places in north west America, where snow lies throughout long winters, and the short, hot summers have little or no rain. As the scientific name implies, it is a magnificent tree, often 36 to 45m (120 to 150ft) tall. Its short, regularly spaced branches form numerous horizontal strata with clear spaces between each whorl, giving the tree a fine, blue-green columnar symmetry. Only at the top, where the big, upright cones are crowded in great numbers, is this symmetry marred. The dark red bark (hence the name) of mature trees is thick, hard and rough, with deep furrows and rounded ridges. In young trees, which tend to be more conical in shape than their elders, the bark is smooth with chalky white patches here and there, and dotted about with resin blisters. (See illustration facing p.81.)

Distribution and habitat

The natural range of the Red Fir extends from the Cascade Mountains in Oregon (43°N) south over the mountain ranges of northern California and along the western slopes of the Sierra Nevada to the divide between the White and the Kern Rivers. In the northern parts it grows at elevations between 1,520 and 2,140m (5,000 and 7,000ft) and in the southern areas at 1,830 to 2,750m (6,000 to 9,000ft). It is often found where the snow lies thick beneath the trees until as late as July each year, giving the equivalent of 750 to 1,500mm (30 to 60in) of rain, and where the hot, dry summers rarely produce more than 25mm (1in).

These firs do best on steep slopes and on flat areas where snow water does not lie, and they prefer the soils of glacial moraines or the deep soils of unglaciated areas. On such sites, where they are semi-dominant, and at the higher elevations, they grow in mixture with Jeffrey's Pine *Pinus jeffreyi*, Western White Pine *P. monticola*, Lodgepole Pine *P. contorta* and Mountain Hemlock *Tsuga mertensiana*; and lower down with White Fir *Abies concolor*, Sugar Pine *Pinus lambertiana* and sometimes Douglas Fir *Pseudotsuga menziesii*. But on some specially favoured sites they are found in dense, pure stands, even aged with crowns fully closed and little or nothing growing beneath them. In clearings though, such plants as *Ceanothus cordulatus*, Bitter Cherry *Prunus emarginata* and one of the Snowberries, *Symphoricarpus rotundifolius* may occur.

Life history

Good seed years occur every two or three years when big, barrel-shaped 22cm (9 in) cones disseminate their contents at anything up to 160,000 winged seeds per acre (0.4 hectares) of cleared ground. Despite a generally low viability, they frequently germinate on the snow. The seedlings grow very slowly, taking five to ten years to reach a height of 30cm (1ft) and until they top 4.5m (15ft) or so, the terminal shoot rarely grows more than 5cm (2in) a year; after this, growth accelerates.

The tallest known Red Fir in America, according to present records, is 56.7m × 3m (186ft × 10ft), and the largest 54m × 4.5m (177ft × 15ft).

These firs are moderately shade-tolerant; in some areas, where the Lodgepole Pine is the main competitor, they eventually oust the pines.

In the north of the range, in Oregon, a variety *shastensis*, which has smaller cones than the type with long, protruding bracts, is more common. Opinions differ as to whether it is a distinguishable variety since it tends to be more like the Noble Fir *Abies procera* and could be the result of natural hybridisation. In the south of the range hybrids with White Fir *Abies concolor* are suspected.

Timber

The timber of the Red Fir has a reddish tinge and is soft and rather brittle, but with a straight, fine grain. It is of little importance in the timber market, being used in a small way for bridge flooring, mine timber and packing cases. Bark blisters exude Canada Balsam which has medicinal uses as well as being suitable for making certain varnishes.

Introduction, performance and use as an exotic

The Red Fir was introduced to Britain by John Jeffrey in 1851. At the time he thought it was *Abies amabilis*, and it was distributed as such to members of the Scottish Oregon Association. William Lobb repeated the mistake a year later, but the plants from these seeds were discovered by Veitch to be different and the error was rectified.

The tree seems to have done better in Scotland than elsewhere. The tallest in Britain was a specimen at Dunkeld House, Perthshire, which Mitchell measured as 35m × 2.7m (115ft × 9ft 5in) in 1961; by 1970 it was 37.5m × 3m (123ft × 10ft 5in). In 1981 it was measured again and was found to have grown to 40.2m × 3.6m (132ft × 11ft 10in) but even by 1974 it had lost its height record to a tree at Cragside in Northumberland. Indeed, all other Red Firs which have been measured and exceed 30m (100ft) in height, are in Northumberland or Scotland, as some recent measurements show:-

Blair Castle, Perth, 37m × 4.8m (121ft × 16ft 2in) 1987
Cragside, Northumberland, 41m × 3m (135ft × 10ft 2in) 1974
Dunkeld House, Perth, 39.6m × 3.6m (130ft × 12ft 2in) 1987
Glenapp, Ayrshire, 31m × 2.1m (102ft × 7ft 2in) 1987

Malcolm Dunn, in a paper read at the Conifer Conference in 1891, stated 'The Red Fir is in truth a stately tree and one of the handsomest of all the taller-growing conifers for ornamental purposes. It is one of the very hardiest of firs, and is seldom affected by spring frosts, and the timber being straight, clean grained, and of good quality, it will no doubt be a useful forest tree.'

This prophesy does not seem to have been fulfilled. Of the 15 trees measured by Elwes in the early 1900's only two remain. One of them, at Rowallane, Co. Down, was 14.3m × 1.8m (47ft × 6ft) in 1906 and when found by Mitchell in 1968, measured 21.3m × 1.8m (70ft × 5ft 11in).

Handsome though it is, the Red Fir is not common and nowadays is rarely planted; perhaps because it is a very slow grower in early life and, as the Americans have found, it is very susceptible to atmospheric pollution.

This native fir is a popular ornamental in the USA in areas where air pollution is not intense; it seems to be hardy in the eastern States on cool, moist sites. A variety, *glauca*, with whitish needles, is also planted.

Cedar of Lebanon

Cedrus libani

This is the best known of the three true cedars. It is a massive, flat-topped tree with a huge bole from which great side branches emerge low down, arching upwards and bending down again towards the ground. They bear, together with the smaller limbs, dark green needles, in whorls, arranged in tiers on huge flat plates which give the broad crown an immense spread. (See illustration facing p.96.)

Distribution and habitat

The natural range of the Cedar of Lebanon stretches from the Lebanon Mountains in the south, through the mountains of Syria northwards into the Taurus Mountains of southern Turkey and thence eastwards almost to the border with Armenia. On Mount Lebanon itself the cedars have been reduced by cutting and grazing to 300 or 400 old trees in a grove growing in moraine on the borders of a stream at about 1,830m (6,000ft). Until recently, when some protection was given against grazing, natural seedlings had constantly been destroyed and no young trees were to be found. Elsewhere in Lebanon and Syria the cedars are isolated in small groves some of which are now being protected and extended.

In the Taurus Mountains the cedars form a considerable proportion of the conifer forest which, here and there, covers the slopes at altitudes between 1,220 and 1,830m (4,000 and 6,000ft). Other associated species are Cicilian Fir *Abies cilicia*, Grecian Juniper *Juniperus excelsa*, Stinking Juniper *J. foetidissima* and Black Pine *Pinus nigra*.

Life history

The female cones, which take two or three years to ripen, are barrel-shaped and about 12cm (5in) tall, sitting upright, amongst the needles. When ripe they disintegrate and shed their winged seeds, leaving behind a spike on a woody base plate.

Height growth of seedlings is slow, and not much quicker in the sapling stage, rarely exceeding 30cm (1ft) a year for the first 70 years, then trailing off. But increase in girth is much more impressive; trees of 70 years or so often girth 3.3m (11ft) and more.

Lebanon Cedars can live to great ages and attain enormous girths.

A tree in a grove on Mount Lebanon was reported to be 12m (40ft) in girth, but most of its companions measured 6 to 7.6m (20 to 25ft). When drawn up in thick forest these cedars can attain considerable heights – in excess of 38m (125ft) – but the massive, open-grown specimens are usually not much more than 27.5m (90ft). Estimates based on ring counts in a huge lower branch of one of the largest trees in Lebanon in the middle of the last century put the oldest specimens at 2,500 years, and some authorities have even mentioned 4,000 years as being a possible life span for these trees.

Timber

It is generally agreed that the timber used by Solomon for the temple in Jerusalem was that of the Cedar of Lebanon; but some authorities think that it might have been Grecian Juniper *Juniperus excelsa*. In any event the cedar timber is now in such short supply that it has no real commercial value, although its pinkish colour and handsome figure, combined with a pleasing fragrance, made it a valuable timber for cabinet work, interior decoration and house building in former times, hence its scarcity now.

Introduction, performance and use as an exotic

Prior to the latter years of the last century the date of the introduction of the Cedar of Lebanon into Britain was in considerable doubt, with the concensus favouring the theory that John Evelyn was the introducer, some time between 1665 and 1670. But in 1879 a very large tree was blown down at Wilton House, Salisbury, which had a girth of 11m (36ft), containing 236 annual rings. Hence this tree must have been planted about 1638, long before Eveyln's time; as there was no reason to doubt the word of the head gardener at Wilton, Mr. Challis, who reported the find, this is now accepted as the date of introduction.

Although Loudon says that the oldest Cedars of Lebanon standing in 1838 were those at Chelsea, planted in 1683, Mitchell noted that the oldest still alive in 1970 was one at Childrey Rectory, near Wantage, planted in 1646. It measured 17m × 7.6m (56ft × 25ft 2in). Elwes does not mention this tree but gives as his choice of the finest specimen the one he photographed at Goodwood, Sussex; one of many hundreds planted by Peter Collinson for the Duke of Richmond in 1761. In 1906 it was 29m × 8m (95ft × 26ft 6in), with a branch spread of 133 paces in circumference. When Mitchell measured this tree in 1987 it was 39.6m × 8.9m (130ft × 29ft 6in).

In 1971, the tallest Lebanon Cedar in Britain was one at Petworth, Sussex, which Mitchell measured as 40.2m × 5.4m (132ft × 18ft); it had maintained its paramountcy since 1905 when Elwes found it to be 38m × 4.4m (125ft × 14ft 6in). It was superseded by a tree at Leaton Knolls, Shropshire, which, in 1981, measured 42.7m × 3.5m (140ft × 11ft 8in). Other recent measurements include the following trees:-

Bowood, Wilts., 41m × 5.7m (135ft × 19ft 4in) 1975
Blenheim Palace, Oxon, 29m × 8m (95ft × 26ft 9in) 1978
Preston House, Middlesex, 26m × 8m (85ft × 26ft 7in) 1988
Stowe, Bucks, 31m × 8m (102ft × 26ft 9in) 1981
(the tree at Petworth House, Sussex, which measured 40.2m × 5.5m (132ft × 18ft 8in) in 1976, was blown down in 1987.

Ever since the Cedar of Lebanon was introduced into Britain it has been a popular ornamental tree and it has been planted extensively in gardens and policies of the great estates as well as in more modest gardens both in town and country. Loudon, Elwes and Mitchell all list large numbers of fine specimens, some massive, short-boled and spreading, others with longer boles and more upright in habit. But whatever their shape these cedars are remarkably impressive, long-lived and tolerant of most soil types except those with a high basic content; and to add to their aesthetic attributes is the mystique of their Biblical connection.

In the USA, where the Lebanon Cedar was introduced in colonial times, and where the same Biblical connections have a special appeal, it is widely grown, especially in the south. It was not hardy in the north until 1903 when the Arnold Arboretum sent a special expedition to Asia Minor. There it collected seed at the highest altitude where cedars grew. The strain from these seeds was called *C. libani stenocoma*. It is hardy in the north east, extending the range of this cedar as an ornamental over most of the country and making it the hardiest of the three true cedars.

Atlas Cedar

Cedrus atlantica

Of the true cedars the Atlas Cedar is the straightest and most upright growing. From the bole, clad in dark grey bark fissured into short, narrow flakes, the main branches emerge at 45° angles to form a broadly columnar crown of grey-green needles. These are held in whorls on spurs in the old wood, the branch ends ascending to give the whole tree an upswept look, with rarely more than one main stem. (See illustration facing p.96.)

Distribution and habitat

The Atlas Cedar, as the name implies, is native to the Atlas Mountains of Morocco and Algeria, and is found there growing on a variety of soils, calcareous and otherwise, at altitudes of between 1,220 and 2,140m (4,000 and 7,000ft). Here the winter rainfall is about 750mm (30in), where snow lies for three months of the year and the summers are hot and dry.

Some of the finest stands are found in the Teniet-el-Haad area in western Algeria and in the Btana area in the east, around the 2,140m (7,000ft) Mount Chelia, the highest in the country. The trees either grow in pure stands, usually of limited extent on rocky outcrops or, more often, in rather open mixed forest in ravines and on the north-facing mountain slopes, together with such species as Algerian Fir *Abies numidica*, Evergreen Oak *Quercus ilex*, Mirbeck's Oak *Q. mirbeckii*, and two species of juniper, the Spanish *Juniperus thurifera* and the Phoenician *J. phoenica*.

Life history

The Atlas Cedar starts to bear cones at about 40 years of age. The pale green female cones, barrel-shaped and about 7cm (3in) tall, are set upright amongst the whorls of needles. They take up to three years to ripen, disintegrate and release their copious crops of winged seeds.

Growth, for a cedar, is rapid, seedlings reaching 15cm (6in) in their first year and 3m (10ft) in seven years when acceleration normally occurs to as much as 90cm (3ft) annual height growth. Mature trees can be as much as 45m (150ft) tall, but 33 to 36m (110 to 120ft) is more usual with girths of 4.5 to 6m (15 to 20ft), although Elwes mentions a tree said by a French forestry official in Algeria to have been 19m (62ft) in girth.

Timber

The brownish-yellow timber is strong, durable and fragrant. It is used for general building, furniture and paving blocks.

Introduction, performance and use as an exotic

The Atlas Cedar was first 'discovered' and recognised as a separate species by Philip Webb when he visited Morocco in 1827. It was introduced into Britain somewhere around 1844; at about the same time, it seems, as the beautiful variety *glauca*. Some of the trees recorded by Elwes and named as the type have turned out to be this variety. For instance, the seeds gathered by Lord Somers at Teniet-el-Haad in 1845 and sown at Eastnor Castle, were of *glauca*. The tallest of them, labelled a type tree by Elwes when he measured it in 1906, was 23.4m × 2.4m (77ft × 8ft). In 1970, Mitchell found it to be 32m × 4m (105ft × 13ft 9in), and by 1981 it had grown to 33.5m × 4.2m (110ft × 14ft 4in).

Of the likely original type trees the tallest and largest was at Bowood, New Forest, which measured 40.2m × 5.7m (132ft × 18ft 10in) in 1968. Other type trees more recently measured are:-

Bodnant, Gwynedd, 35m × 3.7m (115ft × 12ft 5in) 1981
Pitt House, Devon, 25m × 5m (82ft × 17ft) 1975
Powis Castle, Powys, 17m × 3.8m (56ft × 12ft 7in) 1981
Westonbirt, Glos. (Jackson), 32m × 4.7m (105ft × 15ft 6in) 1981

The type tree is not so frequently planted as the blue variety of which there are many very fine specimens up and down the country, the tallest, in 1970, being one at Cuffnels, Hants., which measured 34.4m × 2.2m (113ft × 7ft 7in). An original, planted at Dropmore, Bucks, in 1846, was 27.4m × 4m (90ft × 13ft 5in) in the same year. The tree at Murthly Castle, Perth, which was 31m × 4.8m (102ft × 15ft 10in) in 1981, had grown to 32.6m × 5m (107ft × 17ft) by 1988.

Cedar of Lebanon, Polesdon Lacey, Surrey (p.93)

Atlas Cedar, Polesdon Lacey, Surrey (p.95)

Deodar, Westonbirt, Glos. (p.97)

European Larch, Grizedale Forest, Cumbria (p.100)

Morinda Spruce, Penrhyn, N. Wales (p.107)

Golden Larch, Drayton Manor Pk., Staffs. (p,105)

Some recently measured trees of the blue variety include:-

Bodnant, Gwynedd, 22m × 4.5m (72ft × 15ft), 36m × 3.3m (118ft × 10ft 11in) 1981
Eastnor Castle, Herefords., 30m × 4.5m (100ft × 14ft 10in) 1987
Merton Park, Norfolk, 30m × 4m (95ft × 13ft 6in) 1981

This unusually beautiful powdery blue conifer is one of the most popular ornamental trees planted in Britain. It can reach a considerable size and makes a magnificent spectacle alongside the darker greens of other conifers providing a backdrop to flowering shrubs and other plants. But its comparatively narrow habit when young and quick growth make it a very satisfactory subject for smaller gardens, too. Furthermore, it will grow on most soil types including those containing lime. It is not usually propagated by seed but by grafting material from specimens with exceptionally blue foliage.

The blue variety is also very popular in the USA where it was introduced before 1840. It is widely planted in both eastern and western states in parks and gardens, particularly in the north where it is much more hardy than the Deodar.

Three cultivars are also grown: 'Aurea' with golden tinged foliage, and usually a tree no more than 15m (50ft) tall; and, much rarer, 'Fastigiata', very columnar and acutely pointed and reaching heights of 15 to 21m (50 to 70ft); and 'Argentea' with a definite silvery tinge to the blue needles.

As a forest tree in England the Atlas Cedar has been tried without much success largely because of its slow growth in early years making establishment difficult; but it may have a place on dry, calcareous soils in Southern England.

In France, however, where it was introduced as early as 1734, it has been planted as a forest tree since 1848, and in the south of that country there are crops of all ages up to 100 years old. It seeds profusely and spreads naturally on the dry limestone soils of the southern mountains. Near Carcassonne, for instance, old oak and chestnut coppice was converted to Atlas Cedar woodland by thinning the coppice to allow the cedar regeneration from scattered standards to establish itself. The cedar is also established by broadcast sowing and by planting.

Deodar or Himalayan Cedar

Cedrus deodara

The Deodar is the most elegant of the true cedars and also economically the most important. The longer needles, the pendant tips of the branches and the drooping leading shoot, all mark it as different from the more rigidly growing Atlas and Lebanon species. In youth these characteristics are even more pronounced, with the additional attraction that the foliage has a silvery tinge. (See illustration facing p.96.)

Distribution and habitat

The natural range of the Deodar extends right along the Western Himalayas from the Kuram Valley in the north-west to Kumaon in the south-east, at altitudes varying between 1,220 and 3,050m (4,000 and 10,000ft). Extensive pure stands are rare; more usually these cedars are found in little groups or scattered amongst other species of trees. At the lower levels common associates are Chir Pine *Pinus roxburghii*, Evergreen Oak *Quercus incana* and Indian Horse Chestnut *Aesculus indica*. In the middle altitudes, between 1,830 and 2,440m (6,000 and 8,000ft), where the Deodars grow best, and are most numerous, such trees as Pindrow Fir *Abies pindrow*, Himalayan Cypress *Cupressus torulosa* and Bhutan Pine *Pinus griffithii* are found in mixture with them; whilst at the upper limits the Himalayan Fir *Abies spectabilis* may grow with the cedars in sheltered, damper spots with the Deodars growing almost pure, albeit rather stunted, on the dry ridges with, perhaps, a few *Pinus gerardiana*.

After a visit to India in 1978, I wrote the following in *Quarterly Journal of Forestry*:-

> 'The ridge upon which Simla sits and hangs 1,830m (6,000ft) above the sea was once a place where Deodars, Evergreen Oaks and Chir Pines met and mingled in a dense forest, the cedars at the lowest part of their vertical range and the pines at the highest. Everywhere that man has left space enough the cedars still stand: the old ones tall branchless for half their height yet graceful still, their branch tips drooping, and the younger ones clothed to the ground, a silvery sheen to their needles. With other trees they grow amongst the houses, often on such slopes that, with the boles at the bottom of the gardens, the crowns will be level with the upstairs windows.'

Life history

The 12cm (5in) long, barrel-shaped female cones sit upright amongst the whorls of needles. They disintegrate in the autumn, releasing their large, winged seeds and leaving the central cores sticking up from the branches. Good crops of seed are borne every two or three years, but in the dense stands of mixed forest the cedars never cone very heavily, and seed production is not so great as on the sunny ridges where the trees cone much more freely.

Seedling growth is rapid in favourable conditions where shade and moisture are maintained throughout the hot weather and where the competing trees are pines and firs which are less tolerant of shade than the Deodars.

These cedars grow to great sizes. In the early days of the Indian Empire, when the Imperial Forest Service was established, huge trees 75m (250ft) tall and 10.5m (35ft) in girth were found during surveys; but few of these giants are now left, and heights of 55m (180ft) and girths around 4.5m (15ft) are the biggest one can now expect to see. Given the chance, Deodars live to a considerable age. A fallen tree discovered during early surveys was found to be 900 years old, and a section of a Deodar at the Indian Forest Research Station, Dehra Dun, shows 664 annual rings.

This cedar is an important forest tree in India and Pakistan, and it is managed for timber production along the whole of its range.

Timber

The light brown, very durable timber is one of the most important grown in India. It is used for railway sleepers, building and for bridges. It splits well into boards and shingles.

Introduction, performance and use as an exotic

The Deodar appears to have been first introduced into Britain by the Hon L. Melville, who sent seeds home in 1831. Some of the resulting seedlings were grown at Melville House, Fife, and they were still there in 1935, but were felled that year without any record being taken of their measurements.

Loudon, writing in 1838, mentions a Deodar in the Horticultural Society's garden at Chiswick which was seven years old and 2.4m (8ft) tall. It could have been from the same seed as the Melville House trees. Loudon also mentions a tree at Dropmore, Bucks., which he says was from seed sown in March, 1831, and planted out in the autumn of the same year. This may well be the tree which Elwes mentions, although he says it was planted in 1834. In 1905 it measured 23.4m × 2.7m (77ft × 8ft 10in). In 1970, Mitchell who, incidentally gives the planting date as 1849, found it to be 30.8m × 3.2m (101ft × 10ft 10in).

There are two other original trees at Dropmore as well as at Westonbirt, all between 24.5 and 26m (80 and 85ft) tall. The two originals at Bicton, Devon, growing in the Italian Garden, one of which was over 30m (100ft) tall, were killed by honey fungus in the early 1970s. A Westonbirt tree measured 37m × 3.6m (121ft × 12ft 4in) in 1981.

In 1841, the Woods and Forests (the forerunner of the Forestry Commission) imported a large quantity of Deodar seed and raised about 60,000 seedlings with which an attempt was made to afforest parts of the New Forest and the Forest of Dean, without much success. A few trees still survive in scattered groups but have never attained anything like the size of some of the north west American conifers by which they were suppressed, as indeed they are by native oaks and other species.

From this second batch of seed and from others imported in the middle of the 19th century, there are many fine specimens still growing:-

Ballandean, Perths., 29.8m × 5m (98ft × 16ft 7in) 1987
Charleville, Co. Wicklow, 25m × 4.5m (85ft × 15ft 2in) 1987
Eastnor Castle, Herefords., 36m × 4.6m (118ft × 15ft 8in) 1988
Minstead Park, Hants., 37m × 3.8m (121ft × 12ft 9in) 1981
Whitfield, Herefords., 36.6m × 4.5m (120ft × 14ft 10in) 1973

As an ornamental, this cedar is an excellent tree, especially when young. Early height growth is very rapid – as much as 90cm (3ft) a year – and the elegant, drooping branches, silvery-green in colour and clothing the tree down to the ground, make it a most attractive subject for a small garden.

As a forest tree, early experiments failed. Later plantings fared little better, although at Bagley Wood, Oxford, a plot of Deodars, planted around 1920, reached a top height of 19m (62ft) and girths of 0.7m (2ft 4in) in 45 years. Even so, no serious attempt has been made to plant it on a commercial scale.

It seems that the Deodar was introduced into the USA at about the same time as to Britain and, although not hardly in the north, it is a very popular ornamental where the climate is suitable. A considerable number of colour variations is offered by the trade in that country.

In Australia and New Zealand, as well as in many tropical and subtropical countries, the Deodar is planted as an ornamental; in such climatic conditions its quicker growth makes the foliage even more attractive.

European Larch

Larix decidua

The European Larch, the only deciduous conifer native to Europe, is a large, essentially alpine tree with a narrow, conic crown of bright green, whorled needles on straw-coloured twigs. These hang from thick, level branches, supported on a straight, tapering stem. The grey bark has fissures exposing a reddish cortex beneath. The lower branches of old trees often form massive upturned appendages. (See illustration facing p.97.)

Distribution and habitat

This larch is found throughout the mountains of Central Europe, ranging from the Maritime Alps in France, eastwards through Switzerland, Austria and Southern Germany and on to the Tatra Mountains and the Carpathians. Elwes says that the highest recorded elevation at which the European Larch grows is 2,500m (8,200ft); it is rarely found lower down than 430m (1,400ft) except in parts of Poland and Czechoslovakia, particularly in the upper valleys of the Vistula and the Odra Rivers, where it may be found slightly below this altitude.

At its upper limits this larch consorts with Arolla Pine *Pinus cembra* and Mountain Pine *P. unciata*, whilst lower down it is frequently found in mixture with Beech *Fagus sylvatica* and Silver Fir *Abies alba*.

Life history

The beautiful, pink female cones appear in profusion on the two-year-old wood and in lesser numbers on the older wood, gradually ripening to dark brown with the tips of the scales turned outwards. The seeds are shed during the winter and spring, but cones may remain on the trees for five years or more.

The growth of seedlings is very rapid – up to 38cm (15in) in the first year, with annual shoots up to 1.2m (4ft) long in the immediate following years. Mitchell records a Sudetan Larch which was 1.2m (4ft) tall when planted; six years later it had grown to 7.3m (24ft), and to 14m (46ft) 16 years after that. He also recorded a larch at Dunkeld which measured 33.8m (111ft) at the age of 82 years.

Elwes reports that the largest larch in the Alps which he saw was a tree near Modane in the Forest of Villaroden, at an elevation of 1,370m (4,500ft), measuring 27.4m × 4.8m (90ft × 16ft). He gives a photograph of what was reputed to be the finest specimen in the Alps, at Blitzlingen in Upper Valais, at an elevation of 1,340m (4,400ft). Although its top was dead, and it thus measured only 29m (95ft) in height, it girthed 7.4m (24ft 6in), and its lower branches, one of which was characteristically immense, stretched 9m (30ft) from the bole on either side.

Although, on account of the value of their timber, larches are rarely allowed to stand for more than 200 or 300 years, some sawn stumps in the Alps have shown over 400 rings, and trees of over 700 years have been recorded.

Timber

European Larch timber is reddish brown to brick-red in colour, with the contrasting lighter sapwood and the darker summer wood giving it the characteristic grain. It has many uses from pitwood, stakes and transmission poles to boat planks and skins, as well as for fencing, gates and other estate uses where moderate durability is an advantage.

Introduction, performance and use as an exotic

The exact date of the European Larch's introduction into Britain is unknown, but Parkinson, in his *Paradisus*, published in 1629, mentions this tree as being planted but rare; and Evelyn, in 1664, mentioned a larch tree of goodly stature growing in Chelmsford, Essex.

Some of the first larches of known dates were nine trees planted at

Dawyck, near Peebles, in 1725. Loudon gives a drawing of one of them which, around 1834, measured 24.4m × 4.5m (80ft × 15ft) at base. Others were planted at Dunkeld in 1727; and between 1740 and 1750, James, Duke of Atholl, planted 350 larches at Dunkeld and a further 873 at Blair Athol. In 1788 the Society of Arts offered three gold medals for planting larch and making known the useful properties of its timber. This publicity started a trend during the 19th century when more larch were planted than any other tree, including oak.

Headrick, in his *History of the County of Lanark*, published in 1813, mentions three larches which were planted at Lee Park in 1675. Mitchell found them all in 1954, but could only locate two in 1971 when they measured 24.4m × 3.9m (80ft × 12ft 11in) and 21.3m × 3m (70ft × 10ft 5in). In 1954, the third, and now missing tree, measured 25m × 3.2m (82ft × 10ft 6in).

Both Loudon and Elwes give much prominence to the two larches at Dunkeld, planted in the cathedral grounds in 1727. In 1831, the larger measured 30m × 3.2m (100ft × 10ft 6in). It was the same height but girthed 4.7m (15ft 8in) when Elwes measured it in 1904. By 1971 only one still stood, and it is still there. Mitchell found it to be 32m × 5.2m (105ft × 17ft 4in) in 1981.

Seedlings from these two larches were planted at Kennel Bank, Dunkeld, in 1750 and have become the magnificent specimens seen there today. In 1971, one of them was 40m × 3.2m (131ft × 10ft 8in), with a clean bole of 27.4m (90ft). In 1885, the then Duke of Atholl planted some Japanese Larches below Kennel Bank as an eventual seed source of this recently introduced species. In 1904, some seedlings beneath them were seen to have very vigorous but pale shoots, unlike the Japanese species' pink ones. They turned out to be the first Hybrid Larches *Larix* x *eurolepis*.

There are many magnificent European Larches in Britain, especially in Scotland. Recent measurements include some of them:-

Ardvorlich, Perths., 39.6m × 4.8m (130ft × 15ft 7in), 42m × 2.7m (138ft × 9ft 2in) (planted, 1780)
Blair Castle, Perth, 36.6m × 4.5m (120ft × 14ft 11in), 42.7m × 4.3m (140ft × 14ft 6in) (planted, 1737) 1988
Bonskeid, Perth, 33.8m × 5m (111ft × 16ft 6in) 1988
Durris House, Kincards., 22m × 4.7m (72ft × 15ft 8in) 1988
Nonsuch House, Bronham, Wilts., 26m × 4.8m (85ft × 16ft) 1979
Pothill, Auchterarder, 15.8m × 5.4m (52ft × 17ft 9in) 1988

As a forest tree the European Larch is still much planted in Britain, particularly on private estates where its timber is a valuable and useful asset. Both the Hybrid and the Japanese species produce a greater increment, but seed scarcity of the former and the tendency to crooked growth in the latter ensure a good future for the European. If the provenance is correctly chosen to avoid canker and if properly thinned, the European Larch will produce good and beautiful plantations which are a great asset to the countryside.

In the USA, where the European Larch was introduced in colonial times, it is planted more than any other larch, especially in northern areas. A variety 'Pendula' is particularly popular, and the species' habit of retaining its cones for several years is admired.

Japanese Larch

Larix kaempferi

The Japanese Larch is a large, deciduous, sub-alpine tree indigenous exclusively to Japan. The soft, light green needles, borne in clusters on spur shoots, on horizontal branches, like the true cedars, form a conical crown. This is supported on a bole clad in reddish-brown, scaly bark. When leafless, the twigs can be seen to be orange-red in colour, as opposed to the straw-colour of the European species.

Distribution and habitat

The natural range of the Japanese Larch is confined to Central Honshu where it grows on slopes of the volcanic mountains at altitudes of 1,220 to 1,830m (4,000 to 6,000ft), and in places higher still, as on Fujiama, where it is found at over 2,750 (9,000ft). At the higher levels, near the timber line, it grows in mixture with Maries and Veitch's Firs *Abies mariesii* and *A. veitchii*, Hondo Spruce *Picea jezoensis* var. *hondoensis* and Japanese Hemlock *Tsuga diversifolia*; and lower down with oaks, maples and other hardwoods.

Life history

The pink female cones open in April/May, turning brown as they ripen in October and shed their seeds from their down-turned, rose-petal-shaped scales.

Seedling growth is rapid after a rather slow and branchy start, but later height growth can be as much as 90 to 120cm (3 to 4ft) a year. Mature trees in Japan can be up to 36m (120ft) tall and 3m (10ft) in girth, but 21 to 24m (70 to 80ft) is more usual over most of the mountainous range.

This larch is extensively planted throughout Japan as a forest tree and, in consequence, it is one of the most important timber producers in the country. It is also a popular ornamental, providing an early spring green as well as yellow autumn tints.

Timber

The reddish-brown, rather resinous timber is extensively used in Japan for construction work, mining timber and many general uses.

Introduction, performance and use as an exotic

The Japanese Larch was introduced into Britain in 1861, when J. Gould Veitch sent seeds home to his Exeter nursery. It seems that very few plants were raised from these, as only three of these originals now exist. One, at Tortworth, Glos. (now Leyhill Open Prison), planted there in the early 1860's, was photographed and measured by Elwes in 1904 when he found it to be 13.7m × 1.3m (45ft × 4ft 7in). Sixty years later, when Mitchell measured it, it had only managed to grow to 17.6m × 1.6m (58ft × 5ft 7in). This poor performance, noted over the years, gave this larch a quite unjustified reputation. Another original at Kew did not help much to put the record straight: it managed only 19.2 × 1.2m (63ft × 4ft) in 103 years. Possibly the provenance of the first batch of seed was at fault, for later plantings from different seed sources have done much better. A tree at Blair Atholl, Perthshire, planted in 1886, was 40.2m × 2.9m (132ft × 9ft 9in) in 1987, and perhaps the tallest in Britain. Other notable trees include the following:-

Avondale, Co. Wicklow, 26m × 2.4m (85ft × 8ft 1in) 1980
Blairadam, Fife, 18.9m × 2.8m (62ft × 9ft 6in) 1987
Bonskeid, Perth, 29m × 2.7m (95ft × 9ft 3in) 1987
Glen House, Peebles, 29.8m × 3m (98ft × 9ft 10in) 1987
Glenapp, Ayrshire, 31m × 2.3m (102ft × 7ft 9in) 1987
Heanton, Devon, 26.8m × 1.9 (88ft × 6ft 8in) 1981

The Japanese Larch is an important forest tree in Britain. It is an excellent pioneer species on better soils and in higher rainfall areas. It grows with great rapidity, producing a valuable timber for general work. It can even be established on spoil tips and on other degraded land, quickly producing ground cover and a build-up of humus from its copious production of deciduous needles. It is also more resistant to larch canker than the European species. However, because of its relatively demanding nature as regards soil quality and rainfall, it may be superseded by a hybrid between it and the European, *Larix* x *eurolepis*, which grows quickly in drier soils, and is canker-resistant.

This hybrid occurred naturally for the first time at Dunkeld, Perthshire around 1897 in seed from Japanese Larches in the Avenue there. The earliest recorded hybrid seedlings were planted at Inver, near Dunkeld in 1904. Neither the Japanese Larch nor the hybrid grows as quickly as some of the north western American conifers on similar soils, and they are thus never likely to be planted as extensively.

The Japanese Larch was introduced into the USA at about the same time as into Britain and it is reckoned to be the best ornamental larch in that country where, as in Europe, it is less susceptible to canker.

Golden Larch

Pseudolarix amabilis

The Golden Larch is a deciduous conifer, attaining, in its natural habitat, heights of 30 to 36m (100 to 120ft). The bole has a reddish-brown, scaly bark and supports long, level branches bearing large, coarse larch-like needles. These are held in clusters of 15 to 20 on the ends of long, curved spurs, the whole forming a broadly conic crown. The needles turn a golden yellow in the autumn. (See illustration facing p.97.)

Distribution and habitat

The Golden Larch is found growing naturally in limited areas in the Provinces of Anhwei and Kiangsi in eastern China, where it forms a rare component of the essentially deciduous forests at altitudes varying between 460 to 1,220m (1,500 and 4,000ft) in company with such species as Chinese Tulip Tree *Liriodendron chinensis*, Chinese Sweet Gum *Liquidambar formosana* and a host of other trees, including oaks and limes.

Life history

The male flowers occur on the ends of the spur shoots, and the female cones on short, leafy branches which arise from the spurs. These cones become leathery as they ripen at the end of the first season, with thick triangular scales which gape open to release the winged seeds and then disintegrate to leave only the central stem on the tree.

Once established, the seedlings grow quickly to reach, in their native forests, heights in excess of 30m (100ft). Fortune reported seeing a Golden Larch standing alone and clothed with branches down to the ground, on a mountain slope at 1,220m (4,000ft), which measured 39.6m × 2.4m (130ft × 8ft).

Timber

This is very similar to that of the true larches but it is too rare to be of any economic importance.

Introduction, performance and use as an exotic

The Golden Larch was first discovered by Robert Fortune in 1853, initially as an ornamental pot plant and later growing near a monastery at 460m (1,500ft) south west of Ningpo in eastern China. He sent seeds home in the same year and another batch two years later, but apparently few if any germinated and the only plants living in Britain in 1860 (according to Elwes) were natural seedlings dug up in the Chinese forests and sent home in Wardian Cases in 1854. The first specimen to flower in England was at Lucombe and Prince's Exeter nursery in 1887.

The finest tree which Elwes knew of was at Carclew, Cornwall. In 1902 it measured 10.6m × 1.5m (35ft × 5ft) and, in 1940, 12.2m × 1.5m (40ft × 5ft 2in). It remains the tallest in Britain. In 1962, Mitchell found it to be 19.5m × 2.2m (64ft × 7ft 9in), and in 1988, 22.8m × 2.6m (75ft × 8ft 10in). Another specimen at Scorrier House, Cornwall, planted in 1872, was found by Elwes to be 9m × 0.9m (30ft × 3ft) in 1911 and in 1964, Mitchell measured it as 18.3m × 2.4m (60ft × 7ft 10in). Most of the other Golden Larches in Britain vary between 6 and 12m (20 and 40ft) in height. The Golden Larch planted at the Royal Botanic Garden, Edinburgh in 1908, is the most northerly specimen and was 12.2m × 1.2m (40ft × 4ft) in 1981. Other recent measurements include the following trees:-

Leonardslee, Sussex, 15.2m × 1.9m (50ft × 6ft 6in) 1987
Royal Botanical Gardens, Kew, 17.9m × 1.8m (59ft × 6ft 2in) 1980
Wakehurst Place, Sussex, 12.2m × 1.8m (40ft × 6ft) 1987

Generally speaking, the Golden Larch does not fare well in Britain, as it requires hotter summers than those which normally occur. Thus the best trees are in the south, but nowhere will they thrive on calcareous soils.

In the USA, where hot summers occur more regularly in many parts, the Golden Larch grows well. Elwes noted that the largest specimen he knew of in that country was one at Parson's Nursery at Flushing on Long Island, which had been imported from London in 1859 when 90cm (3ft) tall. In 1895 it was 16.7m (55ft) tall and girthed over 1.8m (6ft), with a crown 15m (50ft) across, and bore seeds regularly. Another specimen at Wellesley, Mass. was 10.6m × 1.2m (35ft × 4ft) in 1905, with a branch spread of 11.2m (37ft). It too had borne seeds since 1887.

Today in the States this tree makes a picturesque and popular ornamental, with its soft, grass-green leaves which turn a beautiful yellow in the fall. But it seems unable to withstand temperatures lower than -23C (-10F) and will not survive in the most northern states. Even in ideal conditions it rarely exceeds 15m (50ft) in height and is on the whole rather slow-growing.

Morinda Spruce

Picea smithiana

The Morinda Spruce is a Himalayan conifer which vies for elegance with the American Brewer Spruce. The 60m (200ft) tall, slender spire-like crown of level branches supports long, hanging shoots covered all round with long, sharp, shiny-green, forward-pointing needles. The bole, a massive 6m (20ft) in girth in old trees, has a greyish-brown bark, split up by narrow fissures and rounded scales. (See illustration facing p.97.)

Distribution and habitat

The Morinda Spruce occurs naturally throughout the Western Himalayas from Kashmir to Nepal, and westward into Gilgit, Chitral and Afghanistan. It grows mostly at altitudes between 2,140 and 3,660m (7,000 and 12,000ft) with the best trees in the middle range of 2,290 to 2,760 (7,500 to 9,000ft) where, in mixed coniferous forests, its main associates are Deodar *Cedrus deodara*, Pindrow Fir *Abies pindrow* and Bhutan Pine *Pinus wallichiana*, with the spruce being more abundant than the others on the drier ridges. In these forests, the dense stands, often hung with lichens, are in ravines, interspersed with open grassy meadows bordered by deciduous species such as maples, Himalayan Birch *Betula utilis* and evergreen oaks. A considerable shrub layer of many species including dwarf bamboos, raspberries and balsams, as well as rich ground flora, is regularly flattened by heavy winter snowfalls.

Life history

The cylindrical 15cm (6in) long, pendant female cones, made up of smooth, leathery scales with curved tips, turn purplish-brown as they ripen in the autumn, when their seeds are scattered by the wind. But little natural regeneration takes place in the stands because of the undergrowth mentioned above. A reasonable succession of younger trees occurs only where the foresters prepare the ground.

Growth is fast in established natural stands, girths of 1.8m (6ft) being reached in 125 years. Old trees can be of great sizes; heights of 64.5m (215ft) and girths of 6m (20ft) and more have been recorded.

Timber

The timber of Morinda Spruce, which resembles very much that of other spruces, being yellowish-white in colour and fairly soft, is used locally for building houses and huts, as well as for water troughs and many other uses made necessary by isolation in remote parts.

Introduction, performance and use as an exotic

The Morinda Spruce was introduced into Britain by means of seeds (in cones) sent to the Earl of Hopetoun, at Hopetoun House, near Edinburgh, in 1818 by Dr Govan of Coupar, whose son lived in India. Six plants were raised from them in the same year and grown in pots for two years, after which they were planted out in the arboretum at Hopetoun House. When Loudon saw them in 1837, the larger one was 5.2m (17ft 6in) tall; and when Elwes measured it in 1911, it was 21.3m × 1.8m (70ft × 6ft 2in). In 1971, Mitchell found it to be 27.4m × 3m (90ft × 10ft 5in).

The tallest known to Elwes was at Melbury, Dorset. He photographed it and measured it in 1906 as 26-27.4m × 2.6m (85-90ft × 8ft 10in). By 1970, according to Mitchell, it had grown to 33m × 3.6m (108ft × 12ft 3in); and by 1980 it was 37m × 3.4m (121ft × 11ft 7in). The performance of two trees at opposite ends of the country and measured by both Elwes and Mitchell at half century intervals, is of interest: one at Smeaton House, in Scotland, planted in 1840, was 20.4m × 1.8m (67ft × 6ft 5in) in 1905 and 24.4m × 2.7m (80ft × 9ft 2in) in 1966; and the other, at Pencarrow, Cornwall, planted in 1842, was 17.3m × 1.8m (57ft × 6ft 7in) in 1907, and 27.4m × 2.5m (90ft × 8ft 5in) in 1970.

A specimen at Taymouth Castle, Perthshire, was, in 1974, the tallest in Britain at 37.8m × 3.8m (124ft × 12ft 10in), and another at the same place, although only 33.8m (111ft) tall, girthed 4m (13ft 4in). Other notable specimens are:-

Eastnor Castle, Herefords., 29m × 3.3m (95ft × 11ft 3in), 33.8m × 2.6m (111ft × 8ft 10in) 1981
Fairburn, Ross & Crom., 25m × 4.2m (82ft × 14ft 4in) 1987
Rhinefield House, Hants., 32m × 3m (105ft × 10ft 4in) 1987
Windsor Great Park, 26.8m × 3m (88ft × 10ft 5in) 1987

This spruce is a fine, elegant ornamental tree for a large garden or park, but it is not easy to establish because of susceptibility to damage by spring frosts; probably for this reason it is not often planted in Britain.

In the USA, too, planting of this handsome conifer has been restricted by its tendency to late spring frost damage, but because of its wide habit, more than that of the native spruces, it is a popular tree where it can be grown.

During a visit to New Zealand in 1981, I noted that the genus *Picea* on the whole did not fare well there, but I did see some outstanding specimens of the Morinda Spruce in both North and South Islands, as well as in the botanic gardens in New South Wales, Australia.

Brewer Spruce

Picea brewerana

The Brewer Spruce is a beautiful and rather rare, medium sized tree of the dry mountain ridges and peaks of the Pacific Slope of north west America, where it forms small groves up to the timber line. A multitude of string-like branches hang in tresses from the slightly drooping, evenly spaced branches below the slender top; presumably an adaptation against snow break, but giving an exceptionally elegant and graceful effect. (See illustration facing p.112.)

Distribution and habitat

The natural range of Brewer Spruce is restricted to the Siskiyou and Shasta Mountains on the borders of California and Oregon at elevations of between 1,530 and 2,140m (5,000 and 7,000ft); mostly on ridges and peaks but sometimes at the heads of streams where it forms small parklike stands. The tree seems to be gregarious and is usually found in little groups interspersed with Ponderosa Pine *Pinus ponderosa*, Incense Cedar *Calocedrus decurrens*, Sugar Pine *Pinus lambertiana*, Western White Pine *Pinus monticola* and Mountain Hemlock *Tsuga mertansiana*.

Following a visit to the Klamath National Park in 1979, I wrote in *Quarterly Journal of Forestry*:-

> 'But the most exciting tree we saw was the Brewer Spruce, a rare tree confined in the main to high elevations in the Siskiyou Mountains. Fortunately, there were two specimens on the edge of the forest road near Happy Camp at about 4,000 feet (1,220m) and there the buses stopped to allow a favoured few to alight and take photographs. Being forest grown, and but lately released from competition with Douglas Fir and Incense Cedar, they were not the beautiful trees we see in gardens at home, well furnished with elegant hanging fronds; but the pendulous branchlets could plainly be seen in the crowns, as well as the cones, and their identity was obvious.'

Life history

The purplish-green 10cm (4in) long, female cones hang from the slender branchlets, shedding most of their seeds in the first autumn, but some may remain stuck in the cones until the second autumn.

Seedling growth is exceedingly slow, as indeed is growth later on, although the tree eventually reaches heights of 24 to 30m (80 to 100ft) with girths of 2.4 to 2.7m (8 to 9ft).

Timber

Such a rare tree produces no timber in sufficient quantity to warrant exploitation.

Introduction, performance and use as an exotic

The Brewer Spruce was discovered by Professor W.H. Brewer in 1863 on Mount Shasta, but positive identification was not claimed until 1884 when T.J. Howell found some trees in the Siskiyou Mountains near the headwaters of the Illinois River. In 1891, R. Douglas collected a quantity of seed and from it raised thousands of seedlings in America, but they all died within two years. According to Elwes, plants were, however, raised from some of the same batch of seed sent to Germany, and they were still alive in 1895. Independent of all this, a naturally grown seedling was sent to Kew in 1897; in 1905 it was 75cm (2ft 6in) tall and then the only living specimen in Britain. By 1931 it had grown to 5.4m × 0.4m (18ft × 1ft 6in), and 11m × 0.6m (36ft × 2ft) by 1963. It produced cones for the first time in 1920.

Elwes' somewhat gloomy forecast, in 1905, that this spruce might well become extinct, has happily not been fulfilled, for the tree has not quite so limited a distribution as he at that time thought. Happily too, his other forecast that 'We need not despair of seeing this beautiful tree established in the south of England', has been amply justified. There are many of them in collections up and down the country, and there are few more elegant conifers grown in Britain today, although its slow growth rate makes it less commonly planted than it deserves.

As a seedling it takes four to five years to reach a height of 30cm (1ft), after which it may add 10 or 12cm (4 or 5in) a year but not always to the leading shoot; sometimes to the laterals only, when the tree tends to become temporarily misshapen.

Most of the older specimens planted in the early years of this century have put on an average of 20cm (8in) a year. Mitchell measured 30 or so specimens in 1963 and none had exceeded this rate, the tallest, 14.6m (48ft), was planted at Vernon Holme, Kent, in 1908; in 1931 it was 6m (20ft) tall. But once established and growing steadily no tree is a better ambassador for its fellow conifers than this strikingly handsome spruce.

Some recent measurements include the following trees:-

Dawyck, Peebles (early introduction), 16m × 1.8m (52ft × 6ft) 1974
Glentanar, Aberdeen, 17m × 1.8m (56ft × 6ft 6in) 1988
Scone Palace, Perth, 13.4m × 1.1m (44ft × 3ft 8in) 1981
Thorpe Perrow, Yorks., 14.6m × 0.8m (48ft × 2ft 10in) 1981
Westonbirt, Glos., 12.5m × 1.2m (41ft × 4ft 1in) 1981
Wisley, Surrey, 18m × 1.2m (59ft × 4ft 3in) 1988

In the USA this native spruce seems to be very difficult to establish when planted outside its natural range in the north west; but where it will grow it is a very popular ornamental.

Colorado or Blue Spruce

Picea pungens

The Blue Spruce is a tree of a very limited area high up in the Central Rockies. Its sharp, stiff, blue-green, sometimes even silvery-white, needles in the narrow, conic crown make it a particularly arresting sight. (See illustration facing p.112.)

Distribution and habitat

Whilst this spruce is found as far north as Glacier National Park in northern Montana, its main distribution is in Colorado, eastern Utah, northern central Arizona and eastern New Mexico; and usually at altitudes of 1,830 to 2,590m (6,000 to 8,500ft), and sometimes as high up as 3,050m (10,000ft).

This spruce is rarely found in pure stands, but usually as individuals or in scattered groves along stream banks where it consorts with Douglas Fir *Pseudotsuga menziesii*, Englemann's Spruce *Picea englemannii*, Alpine Fir *Abies lasiocarpa* and Narrow-leaved Cottonwood *Populus angustifolia*, mostly on dry, gravelly soils.

Life history

The 7cm (3in) long, chestnut-coloured, pendulous female cones discharge their seeds in the autumn. The seedlings are slow to grow as indeed is the tree in later life taking, in its mountain stronghold, anything from 200 to 300 years to reach a height of 21 to 27.5m (70 to 90ft). Sometimes, on very favourable sites heights of 36 to 39m (120 to 130ft) may be reached. As is so often the case with trees, slow growers are long-lived, and the Colorado Spruce is no exception; ages of up to 600 years have been recorded.

Timber

The timber is of no commercial importance but may be used locally for poles, posts and for firewood.

Introduction, performance and use as an exotic

This spruce was discovered in 1862 on Pike's Peak in Colorado by Dr C.C. Parry who sent seeds to Harvard University. One plant from these grew to a height of 4.8m (16ft) in Professor Sargent's Massachusetts garden. Waterers Nurseries took a cutting from this tree, which was a particularly glaucous specimen, and planted it in 1877. Most of the larger Colorado Spruces growing in Elwes' time were probably cuttings from this cutting, grafted onto Norway Spruce stock. Elwes did not measure the original cutting at Waterers but, in 1931, it was 12.5m × 1.2m (41ft × 4ft), and when Mitchell measured it in 1961 it was 19m × 1.6m (62ft × 5ft 5in). By 1979 it had grown to 21.3m × 1.8m (70ft × 6ft 1in). A specimen at Highnam, when measured by Mitchell in 1970, was 22m × 1.2m (72ft × 4ft) and is probably the same cone-bearing one that Elwes measured as 9m (30ft) in 1910. The type tree has not often been planted in Britain but there are some trees of note including the following:-

Bedgebury, Kent, 14m × 0.7m (46ft × 2ft 4in) 1981
Hergest Croft, Herefords., 17m × 1.2m (56ft × 4ft) 1980
Leonardslea, Sussex, 19.5m × 1.1m (64ft × 3ft 9in) 1979

All the other trees mentioned above, as well as many others measured by Mitchell and varying between 20 and 26m (65 and 85ft) tall are what are loosely called the Blue Spruce *Picea pungens* 'Glauca' – a tree which is now very common in parks and gardens in Britain. The young shoots in particular are intensely grey-blue and, although these spruces when older tend to become rather scruffy as the needles thin out in the lower branches, they are without doubt amongst the most beautiful of ornamental conifers; and their slow growth is no disadvantage in the average sized garden.

Some recent measurements of this blue variety include the following areas:-

Batsford Park, Glos., 19m × 1.9m (62ft × 6ft 8in); 21.3m × 1.8m (70ft × 6ft 1in) 1980
Radley College, Berks., 25m × 1.6m (82ft × 5ft 7in) 1981
Wimpole Hall, Cambs., 20m × 2m (66ft × 7ft 1in) 1981

An even bluer cultivar 'Koster' which tends to spread out rather than grow up, makes a most attractive subject on a rockery.

In the USA, this native spruce is a very popular garden tree but there, as elsewhere, it does not grow old gracefully; the lower branches die off and remain attached to the bole in a most unsightly manner. Many other cultivars are sold by the trade in addition to 'Glauca', including 'Argentea', which is silvery white; 'Moerheim' which is dense growing and very blue; and 'Pendens' which seems to be another name for 'Koster'.

The Blue Spruce is extensively planted in New Zealand and in parts of eastern Australia and Tasmania as a garden tree.

Brewer Spruce, Queenswood, Herefords. (p.108)

Colorado Spruce, Scone Palace, Perths. (p.111)

Norway Spruce, Mürren, Switzerland, with Mt Eiger in background (p.113)

Oriental Spruce, Amache Arboretum (p.116)

Sitka Spruce, in rain forest, Olympic Peninsula, Washington (p.117)

Serbian Spruce, Bedgebury Pinetum, Kent (p.120)

Western Hemlock, Cragside, Northumberland (p.122)

Western Hemlock, Hutton-in-the-Forest, Cumbria (p.122)

Norway Spruce

Picea abies

The Norway Spruce, the traditional Christmas Tree of Europe, is, when open grown, a tree of classic coniferous symmetry. It has a regular conical crown of horizontal branches at lower levels, somewhat upswept nearer the top, bearing branchlets either side with dark green, short stiff needles, parted beneath and pointing forwards. The bark of young trees is thin, reddish-brown and somewhat papery, but in maturity it becomes darker in colour and cracks into rounded plates. (See illustration facing p.112.)

Distribution and habitat

This spruce is a native of Europe (but not of the British Isles) its distribution extending from the Pyrennees eastwards to the Alps and the Balkans, northwards through Germany to Scandinavia and eastwards to the Carpathians, and on into western Russia. It is mostly a tree of higher altitudes. Some of the best natural stands still survive in the French Vosges and in the Hartz Mountains of Germany at altitudes varying between 610 and 1,220m (2,000 and 4,000ft). A common associate of this spruce in France and Southern Germany is the Silver Fir *Abies alba* with the addition of some Beech *Fagus sylvatica* at lower levels in the Hartz, the Carpathians and in the Balkans. On the North German Plain and on to Scandinavia, where forests occur at much lower levels, common associates are Scots Pine *Pinus sylvestris*, Birch *Betula pendula*, and Alder *Alnus glutinosa* as well as some willows *Salix* spp.

In Norway itself, where the main conifer was Scots Pine, this spruce, despite its name, is thought to have been a recent immigrant from Sweden and Lapland, as there are no remains of it in Norwegian peat mosses. In the northern part of Sweden and western Russia *Picea abies* is replaced by the Siberian Spruce *P. obovata* and a wide zone of hybrids is to be found in between the main distribution areas of the two species.

Life history

The male flowers develop in the spring from the axils of the needles on the branchlets and the female cones appear on the tips of these branchlets. Erect at first but gradually becoming pendulous and changing from green to brown in the autumn, the cones can be anything up to 15cm (6in) long – the longest cones of all the spruces. The winged seeds fall in late winter and early spring and take about five weeks to germinate in good conditions. The seedlings reach about 10cm (4in) in height after the first season. Thereafter growth can be rapid on favourable sites with leaders up to 90cm (3ft) long in the early years but decreasing later to give a maximum height of about 45m (150ft) an girths of 3 to 3.6m (10 to 12ft). However, in Central Europe much taller and larger trees have been recorded. Elwes mentions a tree in the Carpathians which was reputed to be 69m × 3.3m (226ft × 11ft). In Scandinavia, however, such sizes are never reached and the tallest recorded in Norway was a specimen in Hurdalen (Lat. 60°24'N) which was 39.6m (130ft) tall, and in Sweden a tree in South Rendalen was 44.5m (146ft) tall. In 1976, I photographed a tree in the Vosges which was 54m (177ft) tall.

Norway Spruce seems to live longer than its chief companion, the Silver Fir. Elwes states that in Bohmerwald in Bohemia many of the 36 to 45m (120 to 150ft) tall spruce growing at an altitude of between 920 and 1,220m (3,000 and 4,000ft) showed as many as 700 annual rings and others at slightly higher levels exceeded 400 rings. But these were probably exceptional, and over its natural range, as a whole, a life span of 300 or 400 years seems more likely. A tree felled in Tarcau, Romania in 1888 was 392 years old and 59m (195ft) tall. Elwes mentions a 52m (170ft) tree, 3.6m (12ft) in girth, which, when felled in the Capella Mountains of Croatia in 1910, proved to be 300 years old.

Timber

The timber of this spruce, known as whitewood in the trade, varies in colour from almost white to a yellowish brown, and has a natural lustre. It seasons well with little tendency to warp or check. Larger sizes are used for general joinery, carpentry and indoor finishing, as well as for cheap furniture, pit props, flagpoles etc. Some of the very slow grown trees (12 rings to 2.5cm (1in)), found high in the mountains of Central Europe, produce what is called 'Romanian Pine' or 'Bois de Resonance' used in pianos and the bellies of violins. The timber is also used extensively in the manufacture of paper and pulp.

Introduction, performance and use as an exotic

The exact date of the introduction of the Norway Spruce into Britain is now known, but it seems likely that it was brought here in the early 16th century. Turner includes it in his *Name of Herbs*, published in 1548, and both Gerard and Parkinson found it in different parts of the country.

Although this spruce has been growing in Britain for centuries, none has ever lived to any great age. Loudon says that the largest specimen he knew of was a tree 40.2m (132ft) tall at Studley Royal, which he saw in 1906 when he was told it had been planted around 1750. Elwes photographed and measured a tree at the same place (presumably the same tree) in 1905 when it was 42.7m × 3.8m (140ft × 12ft 10in), free of branches for 9 to 12m (30 to 40ft) and apparently in good health. At the time of the Conifer Conference in 1931, this same tree measured 45m × 4.2m (150ft × 14ft 1in). In 1956, Maynard Greville measured a spruce at Studley Royal (again presumably the same one) which he found to be 47.5m × 4.2m (156ft × 14ft 2in); but when Mitchell searched for it in

1966, he could not find it. It had fallen or been felled at the age of about 200 years. This seems to have been the only Norway Spruce in Britain with an authentic, recorded life history. None of the trees recorded by Loudon can be found, nor is their fate known. None of the 11 Elwes trees has survived either, and only five of the 40 specimens mentioned in the 1931 Conifer Conference report can now be found.

The tallest which Elwes knew of was one at Rooksbury Park, Hants., which topped 45.4m (149ft), but recent measurements include one or two which beat that:-

Aira Force, Cumbria, 32.6m × 3.8m (107ft × 12ft 9in) 1976
Cawdor Castle, Nairn, 37.8m × 3.7m (124ft × 12ft 5in) 1980
Fairburn, Ross & Crom., 37m × 3.7m (121ft × 12ft 7in) 1988
Glamis Castle, Angus, 41m × 3.8m (135ft × 12ft 9in) 1981
Hackwood Park, Hants., 37m × 3.2m (121ft × 10ft 7in) 1977
Kincardine Castle, 45m × 3.8m (150ft × 12ft 10in) 1977
Lingholm, Cumbria, 36.8m × 4.2m (121ft × 14ft 3in) 1979
Moniac Glen, Inverness, 41m × 2.5m (135ft × 8ft 8in) 1980
Powerscourt, Eire, 42m × 3.3m (138ft × 11ft 4in) 1980
Trebartha, Cornwall, 45m × 2.7m (150ft × 9ft 2in) 1981

As a forest tree in Britain, this spruce has been popular for two centuries or more on private estates, and has become increasingly so with the Forestry Commission in recent years because of its pulping qualities. Although it is not planted over such vast areas as the Sitka Spruce, it is particularly useful where frosts are severe. Its popularity as a Christmas Tree, a custom introduced by the Prince Consort in 1844, has resulted in great numbers of this spruce being planted specially for this trade, despite a current threat from plastics.

The Norway Spruce was introduced into North America in Colonial times and it was extensively planted in the eastern states. Early growth there is extremely vigorous. Elwes saw a tree at Wellesley, Massachusetts, in 1904, which had reached a height of 24m (80ft) in 42 years and had a branch spread of 18m (60ft). His photograph shows it to have had rather sparse foliage in the upper branches but very well clothed in the lower parts, with arching branches touching the ground. But, in general, this spruce does not mature gracefully in America, the top thinning after about 30 years, and Wyman considers it to have been much over-planted in the past as an ornamental. Certainly, Pinchot's prediction at the turn of the century that 'it would be planted on a large scale in the cutover land of the north . . . as it is in every way superior to native spruces' seems to have come to nought.

Norway Spruce has a tendency to produce sports. In Britain commonly planted cultivars are 'Aurea' which is yellow in spring and early summer; 'Pendula', a weeping version; and 'Virgata', a vigorous and sparsely branched tree. Wyman says that 'Columnaris' and 'Pyramidata' are popular in America. There are also many dwarf and semi-dwarf varieties.

Oriental Spruce

Picea orientalis

The Oriental or Caucasian Spruce has a dense crown when mature (more conic when young) of glossy, dark green, very short needles (the shortest of all spruces) closely pressed to the shoots. The dense branches are supported by a considerable bole which has a pinkish-brown bark, fissured irregularly into thin, peeling plates. (See illustration facing p.112.)

Distribution and habitat

This spruce is a native of mountainous areas in Asia Minor and the Caucasus, and it is usually found growing at altitudes between 920 and 2,140m (3,000 and 7,000ft). In Turkey, it is most numerous in the mountains between Trabzon on the Black Sea and Erzurum to the south east, where it was first discovered by Tournefort in the early 18th century. Its range stretches eastwards into the Caucasus Mountains and to Georgia, but does not appear to extend further than the 45th meridian. It is usually associated with the Caucasian Fir *Abies nordmanniana*, but often grows at higher levels than this fir, forming the tree-line at around 2,290m (7,500ft).

Life history

The 10cm (4in) cones are borne in profusion on the uppermost branches of the tree. Seedlings grow slowly at first, but after five or six years speed up to an average of 60cm (2ft) or so per annum and continue thus until a height of 18 to 21m (60 to 70ft) is reached, when growth again becomes slow. The tallest Oriental Spruce which Elwes knew of was one felled in Turkey in 1899, when it was 56m (184ft) in height and in excess of 3.6m (12ft) in girth. It was 390 years old.

Timber

There is no indication that the timber of the Oriental Spruce differs in either form or use from that of the Norway Spruce.

Introduction, performance and use as an exotic

The Oriental Spruce was introduced into Europe by Beissner in 1837. Loudon says that it was not in cultivation in Britain in 1838, and it seems to have been introduced here the following year. Elwes extols its virtues as an ornamental and gives details of a number of specimens. One at Penrhyn, which was 17.6m (58ft) tall in 1891, was measured by Elwes in 1906 and found to be 22.8m (75ft) in height. Another at Highnam, which was 20.4m × 2m (67ft × 7ft) in 1905, had grown to 26.5m × 2.8m (87ft × 9ft 7in) when Mitchell measured it in 1970. A specimen at Stanage, Herefords., which was grown from the original 1839 seed, measured 28.3m × 3.2m (93ft × 10ft 7in) in 1970.

As an ornamental tree the Oriental Spruce is excellent, with its attractive foliage and neat, dense habit. There are many fine trees throughout the kingdom as some of Mitchell's more recent measurements show:-

Aigas House, E. Ross, (planted, 1877), 32.6m × 3.3m (107ft × 11ft 3in) 1980
Bowood, Wilts., 33.8m × 3m (111ft × 10ft 2in) 1980
Broxwood Court, Hereford, 27.4m × 3.7m (90ft × 12ft 3in) 1981
Cortachy Castle, Angus, 32.6m × 3.7m (107ft × 12ft 5in); 28m × 3.8m (92ft × 12ft 10in) 1980
Meikleour, Perths., 27.4m × 3.6m (90ft × 12ft) 1988
Scone Palace, Perth, 32m × 3m (105ft × 10ft) 1988
Stanage Park, Powys, 33.8m × 3.2m (111ft × 10ft 10in) 1981
Stanway, Glos., 28.6m × 3.3m (94ft × 11ft 3in) 1981
Stourhead, Wilts., 33m × 3.2m (108ft × 10ft 8in) 1980
Stratfield Saye, Hants., 33m × 3m (108ft × 10ft 1in) 1981

A cultivar, 'Aurea', which has bright pale yellow young growth in the spring, contrasting pleasingly with the dark green older foliage, is sometimes seen. There are three specimens at Bedgebury Pinetum, some 9 to 12m (30 to 40ft) tall.

In the USA, where the Oriental Spruce was introduced in 1850, it is hardy and is considered one of the best spruces for ornamental planting, although it is more susceptible to the spruce bud worm than any other member of the genus. The cultivar 'Aurea' is also planted in that country, as well as a low, broad pyramidal form 'Nana' and one called 'Gowdy' with a very fastigiate habit.

Sitka Spruce

Picea sitchensis

The Sitka Spruce is a magnificent tree, the largest of the spruces and a major species in the great forests of the north west American Pacific coastal strip. The huge buttressed trunk, swollen at the base with protruding roots and often 24m (80ft) or more to the first branch, has a thin, scaly bark and bears a somewhat sparse and open conical crown up to 60m (200ft) in height. The needles, silvery white on their undersides, are stiff, with exceedingly sharp points, and radiate all round the shoots to give an overall blue-grey colour to the crown. Young trees are conical, with long and vigorous leading shoots. (See illustration facing p.113.)

Distribution and habitat

The natural range of the Sitka Spruce is in a long coastal strip, 2,900 kilometres (1,800 miles), closely associated with the 'fog belt' and stretching from Kodiak Island in Alaska to Mendocino County in California, the width varying from a few miles in the extreme south to 200 kilometres (130 miles) in the far north.

The tree reaches its maximum development on the Olympic Peninsula in Washington and on the Queen Charlotte Islands in British Columbia, where the rain forests flourish under the best forest growth conditions in North America.

In the north of the range this tree does not usually grow at altitudes higher than 300m (1,000ft) and nowhere in the range does it go higher up than 920m (3,000ft). Everywhere it likes super-humid conditions and prefers soils rich in organic matter which, in Alaska, means where there is a thick accumulation of raw humus and moss.

The most common associated species is Western Hemlock *Tsuga heterophylla*, and in the north, Mountain Hemlock *T. mertensiana*; whilst further south Pacific Silver Fir *Abies amabilis*, Douglas Fir *Pseudotsuga menziesii*, Western Red Cedar *Thuya plicata* and Redwood *Sequoia sempervirens* also accompany this spruce in varying quantities.

Life history

Growth of seedlings is rapid, although the shallow root system makes it so only where the water supply is constant, those on southern slopes often faring badly. Even those trees growing on moist sites which are exposed find shrub competition so great that they are seriously retarded. Frequently, the only seedlings to do well in such circumstances are those which grow on rotten wood where shrub growth is inhibited. But once established, Sitka Spruce quickly overtops hemlocks and red cedars to become dominant and forms a mixed forest of maximum volume and value. A plot in such a forest in coastal Oregon, 147 years old, contained 76 Sitkas to the acre with an average top height of 63m (210ft) and an average girth of 2.7m (9ft), as well as 13 Western Hemlocks to the acre with a top height of 44m (144ft) and girths of 1.5m (5ft), the whole containing a total volume of 2,350 cu. metres per hectare (34,000 cu.ft per acre). But as even-aged mixed stands of spruce and hemlock reach maturity, the hemlock dies first, since few of them live longer than 500 years, whilst the spruce may go on for up to 800 years and remain as a scattered overstorey beneath which a new crop of the more shade-tolerant hemlock will flourish. The hemlock will eventually dominate the site as the old Sitkas die off or until fire or some other catastrophe clears the site and the spruce can come again.

Sitka Spruce is fairly shade tolerant, more so than Douglas Fir but less so than Western Hemlock and Red Cedar, and pure stands are essentially sub-climax, eventually being replaced by a spruce-hemlock or hemlock-red cedar climax mixture.

Timber

The timber is pale pinkish-brown in colour, soft, straight grained and light in weight. It works easily and planed surfaces have a silky sheen. It is very strong for its weight and is easily kiln dried. In wartime it was in great demand for aeroplane construction, and it is now used for furniture, doors, window frames and pulp.

Introduction, performance and use as an exotic

The Sitka Spruce was discovered in Puget Sound by Archibald Menzies in 1792. David Douglas sent the first seeds home in 1831 and a few plants from these were grown in the Horticultural Society's garden at Chiswick in 1832. According to Loudon, these trees were 90cm (3ft) tall in 1838 and at about this time other plants, mostly cuttings, were available from nurseries at 3 guineas (£3.15) each. But the main batch of early seed was sent by Jeffrey in 1852 and from this most of our present-day large specimens were grown. But some from Douglas' seed, measured by Elwes at the turn of the century and by Mitchell some 65 years later, are still alive. Only one of them has an authentic planting date – a tree at Boconnoc, Cornwall, planted in 1843. Elwes found it to be 24.4m × 4.5m (80ft × 15ft) in 1905, and Mitchell recorded it as 33m × 6.4m (108ft × 21ft) in 1970. Another at Curraghmore, Co. Waterford, was 32.3m × 3.6m (106ft × 12ft) in 1905, 48.8m × 5.9m (160ft × 19ft 9in) in 1968, and 49.4m × 6.3m (162ft × 20ft 10in) in 1974.

Elwes also records a tree at Murraythwaite, Dumfries, planted in 1855, which was 23.7m × 2.6m (78ft × 8ft 10in) in 1904 and which Mitchell measured as 30.8m × 4.2m (101ft × 13ft 11in) in 1966. Mitchell also lists over 70 specimens which Elwes did not mention, including one at Dropmore, Bucks. which was one of the cuttings noted by Loudon and planted in 1841. It measured 22m × 2.4m (72ft × 8ft 2in) in 1891 and 28.3m × 3.7m (93ft × 12ft 8in) in 1970.

Mitchell found that these older trees, although they have grown well (the Curraghmore specimen 23.5m (78ft) in 90 years) have not matched the performance of some later plantings on the best sites, such as flooded valley bottoms in Devon where conditions must be very similar to the natural habitat. Near Dunster, for instance, three trees planted in 1928, measured, after 43 years 36.6m × 3m (120ft × 9ft 10in); 40m × 2.4m (131ft × 7ft 11in); and 37.8m × 2.1m (124ft × 7ft 4in).

Some recent measurements show how well this spruce is growing in Britain:-

Balmacaan, Inverness, 47.5m × 6.7m (156ft × 22ft 3in) 1988
Belladrum, Inverness, 54m × 5.2m (177ft × 17ft 4in) 1988
Castlehill, Devon, 45m × 8m (150ft × 26ft 7in) 1988
Doune House, Inverness, 60m × 5.2m (197ft × 17ft 3in), 59.4m × 4m (195ft × 13ft 8in) 1988
Fairburn, Ross & Crom., 44.8m × 7.6m (147ft × 25ft 5in) 1988

Killiechassie, Perths., 44m × 7m (144ft × 23ft) 1988
Powerscourt, Co. Wicklow, 53.3m × 6.9m (175ft × 22ft 9in) 1988

It is only on ideal sites that the performance of Sitka Spruce is good. An 1866 plantation at an elevation of 275m (900ft) in Roxburghshire, reached a top height of 30 to 36m (100 to 120ft) and 4.8m (16ft) in girth in 90 years. It is because of these performances and the presence of so many large, fine specimens up and down the country, from Cornwall to Caithness, that Sitka Spruce has become the most widely planted exotic forest tree in Britain.

Of all conifer thinnings those of this spruce are the most suitable for pulping.

The Sitka Spruce, native to the north west of the USA, does not grow well in the eastern parts of the country; but where the climate is sufficiently cool and humid young trees with their grey-green foliage make attractive garden subjects, especially the cultivar 'Speciosa' which is slow growing, of more compact habit and with ascending branches.

Serbian Spruce

Picea omorika

The Serbian Spruce is a slender, spire-like tree. The short, dark bluish-green needles are borne on branches which are upswept at the top of the tree, horizontal in the middle, with the lower branches sweeping elegantly downwards, their tips arching upwards. The bole, often swollen at the branch joints, has a brownish-red bark, scaling off in papery pieces. (See illustration facing p.113.)

Distribution and habitat

This spruce has a very limited natural distribution in what is now Yugoslavia, occupying an area little more than 19 kilometres (12 miles) long and 14 kilometres (9 miles) wide in the Drina Valley of south western Serbia. Here it is found growing on limestone rocks at altitudes of between 820 and 1,620m (2,700 and 5,300ft). Pure stands of this relic species, which escaped extinction during the last Ice Age are found only at higher elevations on north-facing cliffs and screes in a very restricted area and in an environment not entirely suited to the species. In such areas, competition from other species is minimal. Lower down it reaches greater sizes but only as a co-dominant with other conifers such as Scots Pine *Pinus sylvestris*, Austrian Pine *P. nigra* var. *nigra*, Norway Spruce *Picea abies*, and Silver Fir *Abies alba*, as well as with Eastern Hornbeam *Carpinus orientalis*, Hop Hornbeam *Ostrya carpinifolia*, Beech *Fagus sylvatica* and Aspen *Populus tremula*.

Life history

The cones, first borne when the tree is 20 to 25 years of age, are most numerous near the top, and ripen a pinkish-brown to drop their tiny seeds (140,000 to 450g (1lb)) in great profusion. Growth of seedlings is vigorous with shoots over 90cm (3ft) long in the second or third season. Later growth slows up to something like 30cm (1ft) a year in early maturity, with ultimate heights around the 30m (100ft) mark. Elwes reported a tree felled in the Drina Valley was 42m (138ft) tall and girthed just over 1.2m (4ft); but this would be an exceptional specimen. It was 137 years old. When visiting this area Elwes found Serbian Spruce growing in clumps on limestone cliffs of a steep-sided ravine, the favourite haunt of chamois, the trees averaging 15 to 18m (50 to 60ft) in height, with some as tall as 27m (90ft).

Timber

The timber does not appear to have any special qualities to distinguish it from other more widely distributed spruces.

Introduction, performance and use as an exotic

The Serbian Spruce was introduced into Britain as seed from Frobert of Zurich in 1884, with a second introduction from Belgrade in 1889. In 1905, Elwes measured three fine specimens at Kew, grown from this later seed, and found the tallest to be 7m × 0.3m (23ft × 1ft 1in); the same size, oddly enough, as that noted in the report of the 1931 Conifer Conference. But Mitchell found it had grown to 19m × 0.8m (62ft × 2ft 9in) by 1970. Most of the early plantings of the last decade of the 19th century and the first decade of the 20th varied between 20 to 27m (65 and 90ft) in height in the 1960's.

Specimens measured during the last few years include the following:-

Endsleigh, Devon, 25.6m × 2.1m (84ft × 7ft 4in) 1981
Headfort, Co. Meath, (planted in 1914), 21.3m × 1.9m (70ft × 6ft 5in) 1980
Leonardslea, Sussex, 25m × 1.2m (82ft × 4ft 2in) 1981
Murthly Castle, Perth, (planted in 1897), 32m × 1.9m (105ft × 6ft 7in); 29.8m × 1.9m (98ft × 6ft 7in) 1981
Savill Gardens, Windsor (planted in 1933), 25.6m × 1.5m (84ft × 5ft) 1981
Sheffield Park, Sussex, (planted in 1910, a very slender, dense column), 29m × 1.7m (95ft × 5ft 9in) 1980
Tregrehan, Cornwall, 29m × 1.5m (95ft × 5ft) 1981

Although the Serbian Spruce is found growing naturally on limestone, it has remarkable tolerance of most soils, even deep acid peats. It does not, though, appear to have any advantages as a forest tree over Sitka or Norway spruces on any major site except perhaps as a late flusher which escapes damage from late spring frosts, and thus can be planted in frost hollows. On good soils its performance is less impressive than Norway Spruce, achieving 8.2m (27ft) top height in 23 years as opposed to 9m (30ft). On poorer soils it made 5.7m (19ft) when the Norway only managed 3m (10ft). So it may have a place as a forest tree and it most certainly has one as an ornamental. Its elegant symmetry and the bluish tinge to its needles which flash white on their undersides when disturbed by the wind make it an attractive garden subject, unlikely to out-grow the site.

Hybrids between the Serbian and Sitka Spruces have occurred naturally at Barcaldine, Argyll; and between Serbian Spruce and Sakhalin Spruce *Picea glehnii* at Murthly Castle, Perth.

A cultivar 'Pendula', a small tree with descending branches and pendulous foliage, is more a curiosity for an arboretum than a pleasing addition to a garden.

In the USA, where the Serbian Spruce was introduced in the 1890's, it is recognised as one of the best spruces for landscape planting. It tends to be almost columnar: an 18m (60ft) tree in the Arnold Arboretum has a branch spread at its base of no more than 4.5m (15ft).

Western Hemlock

Tsuga heterophylla

The Western Hemlock is one of the most beautiful of the great conifers which comprise the Pacific coastal forests of north west America. With a spire-like top to the crown, a drooping leading shoot, even in old age, and branches with pendulous tips covered in short, glossy green needles, gleaming white beneath, all supported on a huge russet brown bowl, the tree has an elegance which its great size does little to diminish. (See illustration facing p.113.)

Distribution and habitat

The natural range of the Western Hemlock extends along the Pacific Coast from the Kinai Peninsula in Alaska in a narrow strip south to the British Columbia/US border and then fans out reaching north western California on the coast and eastwards inland through north eastern Washington, northern Idaho and north western Montana.

The best stands are in the humid coastal regions of British Columbia, Washington and Oregon, and on the lower slopes of the Cascades, in the latter two states at elevations up to 610m (2,000ft). In the northern Rockies, in the east of the range, it may be found as high up as 1,830m (6,000ft).

In the coastal areas the Western Hemlock is found growing in association with Nootka Cypress *Chamaecyparis nootkatensis*, Douglas Fir *Pseudotsuga menziesii*, Sitka Spruce *Picea sitchensis*, Red Cedar *Thuya plicata*, Pacific Silver Fir *Abies amabilis* and Lawson Cypress *Chamaecyparis lawsoniana*; and further inland at higher elevations with Englemann's Spruce *Picea englemannii*, Western Larch *Larix occidentalis* and Mountain Hemlock *Tsuga mertensiana*.

Life history

Western Hemlock is renowned for bearing seeds in great profusion; this starts at the age of 25 to 30 years. The cones, strangely small for so large a tree, open in September or October. The seeds germinate in dense masses beneath the parent trees in the following spring and grow only slowly in the early stages – 15 to 17cm (6 to 7in) in three years – although they tolerate shade for long periods. Once established, the height growth increases and continues at about 60cm (2ft) a year. Under optimum conditions mature trees may grow to 53 to 68m (175 to 225ft) in height and up to 4.5m (15ft) in girth.

Western Hemlock is exceptionally shade-tolerant and will eventually shade out Sitka Spruce and Douglas Fir to claim dominance; but Red Cedar and Pacific Silver Fir, with their greater shade tolerance, often maintain a place in essentially climax hemlock stands.

Timber

The yellowish-brown timber is soft, fine grained and non-resinous. Although not very durable, it is to some extent resistant to insect attack. It is used for all but the heaviest construction work as well as for railway sleepers, boxes, shuttering and pulp. British home-grown timber, although not yet available in sawlog sizes in any great quantity, bids fair to equal that grown in America and is already recognised as a useful general estate timber.

Introduction, performance and use as an exotic

The Western Hemlock was discovered by David Douglas in 1826 but was not introduced into Britain until 1851 when Jeffrey sent seeds home for distribution to members of the Oregon Association.

The tallest specimen which Elwes knew of was at Hafodunos, Denbighshire, which was planted in 1856 and which showed the ability to grow rapidly – a trait which has made the tree so popular. At the age of 12 years it was 8.5m (28ft) tall, and 15 years later was 20m × 1.2m (65ft × 4ft). When measured in 1904 it was 28.6m × 2.4m (94ft × 8ft 5in), and in 1960 Mitchell found it had grown to 36.2m × 3.7m (119ft × 12ft 7in) at just over 100 years of age. However, by 1978 its height had only increased by 30cm (1ft), although it girthed 4.2m (14ft 3in).

A tree at Dropmore, Bucks., planted in 1862, which Elwes photographed and measured, had not done so well; it was only 26.8m × 3.3m (88ft × 10ft 11in) in 1970. Whereas the specimen at Murthly Castle, Perthshire, planted in 1866, and also photographed and measured by Elwes, had grown to 42.3m × 3.8m (139ft × 12ft 8in) in 110 years. The largest of the Murthly trees mentioned below is, perhaps, this tree.

The tallest Western Hemlock, in 1970, according to Mitchell, was at Benmore; it measured 48m × 3.6m (158ft × 12ft), and most of the 20 or so 100-year-old trees in the country at that time exceeded 36m (120ft) in height and 3.3m (11ft) in girth. Recent measurements show some of them to have made considerable growth:-

Bodnant, Gwynedd, 41.5m × 4m (136ft × 13ft 3in) 1981
Doune House, Perth, 45.7m × 6.2m (150ft × 20ft 8in) 1987
Gart House, Perth, 41m × 4.8m (135ft × 16ft) 1987
Glamis Castle, Angus, 44m × 4m (144ft × 13ft 9in) 1981
Lanrick Castle, Perth, 45.7m × 4.3m (150ft × 14ft 6in) 1987
Murthly Castle, Perth, 47.8m × 4.4m (157ft × 14ft 9in), 31m × 4.9m (102ft × 16ft 7in) 1987
Scone Palace, Perth, 41m × 6.2m (135ft × 20ft 9in) 1987

Because Western Hemlock fares so well in Britain and grows quickly into a handsome and essentially wind-firm tree, it has maintained and even enhanced its early popularity as an ornamental.

Ever since the first small plantation of 30 trees was established in Argyll in 1888, this hemlock has been extensively planted as a forest tree in Britain. It reproduces itself readily and is able to bear intense shade until released or transplanted. It thrives and grows quickly on a wide variety of sites and soils, although it does not like excessive exposure or a high pH. Its tolerance of shade makes it an invaluable tree for underplanting, particularly in the conversion of old hardwood sites. It is, however, very susceptible to butt rot caused by honey fungus.

In the USA this native conifer does not do well in the eastern states, presumably because the atmosphere is not sufficiently moist.

Strangely, the Western Hemlock seems to have failed in both Australia and New Zealand, and does not even reproduce itself naturally there.

Japanese Hemlocks

Tsuga diversifolia & T. sieboldii

These two trees, the Northern and Southern Japanese Hemlocks respectively, are very similar in appearance being, in Japan, of moderate size, up to 24m (80ft) tall, picturesque and graceful, with domed crowns of typically short, blunt hemlock needles, irregularly spreading out from the twigs, shiny green above and with whitish bands on the undersides.

Distribution and habitat

The Northern species is confined to the mountains of Central and Northern Honshu, with pockets in the south east of the island. It forms an important component of the sub-alpine coniferous forests of those parts at altitudes between 1,220 and 1,530m (4,000 and 5,000ft) often in pure stands, but sometimes in the higher areas in mixture with Veitch's Fir *Abies veitchii* and Maries Fir *A. mariesii*, and lower down with Nikko Fir *A. homolepis*.

The Southern species meets with the Northern in Central Honshu, but extends south to southern Honshu, Shikoku and Kyushu; and it is rather surprising to find that it also grows wild on the island of Ooryon, east of Korea, but not on the peninsula itself. In Japan it is a dominant tree in the temperate coniferous forests of the southern regions, accompanied by Momi Fir *A. firma* and Tiger Tail Spruce *Picea polita*, and, like them, preferring good moist soils.

Where the two species meet, in central Honshu, intermediate forms are occasionally found.

Life history

The flowers open in June and the small, woody-scaled cones drop their winged seeds in the autumn of the same year.

Seedling growth is very slow but the rate gradually increases and heights of 60m (100ft) and girths of 3.6m (12ft) are eventually reached in trees upwards of 200 years old.

Timber

The pale yellow timber, with a wavy grain, is highly prized for building and interior work, and is said to be superior to that of Hinoki Cypress. The bark is used for tanning fishing nets.

Introduction, performance and use as an exotic

The Southern Japanese Hemlock was introduced into Europe by Siebold in 1850. The seeds of both species were first brought to Britain by J. Gould Veitch in 1861.

Elwes knew of no specimens in excess of 4.5m (15ft) tall in the early 1900's. Mitchell recorded the tallest example of the Northern species, in 1970, as being one at Glamis Castle, which was 15m (49ft). In 1981 it had grown to 19m × 1.2m (62ft × 4ft 2in). Other recently measured *Tsuga diversifolia* include:-

Abercairny, Perths., 18.3m × 2.4m (60ft × 8ft 4in) 1987
Borde Hill, Sussex, 12.5m × 0.7m (41ft × 2ft 5in) 1981
Hergest Croft, Herefords., 15.2m × 1.2m (50ft × 4ft 2in) 1987
Lamellan, Cornwall, 14.6m × 1.3m (48ft × 4ft 7in) 1987

The tallest in 1971 *Tsuga sieboldii* was at Lydhurst, Sussex; it measured

16.4m × 1.3m (54ft × 4ft 6in); but other trees, included below, have now exceeded this height:-

Borde Hill, Sussex, 19m × 1m (62ft × 3ft 3in) 1989
Fota, Co. Cork, 18.3m × 1m (60ft × 3ft 4in) 1987
Grayswood, Surrey, 18.3m × 1.3m (60ft × 4ft 8in) 1987
Ochtertye, Perths., 19.8m × 2.7m (65ft × 9ft); 11m × 2m (36ft × 7ft) 1987
Wakehurst Place, Sussex, (pinetum), 14m × 1.5m (46ft × 5ft) 1989

In the USA, where the Northern species was introduced in 1861, although not as often planted as the three native species, it is gaining in popularity because of its richness of colour and low-growing, rounded habit, as well as being the first hemlock to flush in the spring.

Both species are rare in Britain and are often multi-stemmed.

Douglas Fir

Pseudotsuga menziesii

The Douglas Fir is one of the finest and economically the most important of the great conifers which form vast forests on the North West Pacific seaboard of the United States and Canada. In old stands, which sadly are now very rare, such as the Cathedral Grove on Vancouver Island, the huge trees are up to 75m (250ft) tall with barely room enough to walk between them, the massive boles, hung with lichens, characteristically free of branches for up to 30m (100ft) or more; the volume of timber in each and all of them is enormous. Old trees, open grown, tend to have flat and irregular crowns, the lower branches drooping, the upper ones slightly ascending and the trunks deeply ridged with thick dark brown, corky bark. Young trees form broad, sloping pyramids, the soft dark green needles adding grace to their form; and their smooth, shiny grey-brown bark is marked with resin blisters. (See illustration facing p.128.)

Distribution and habitat

The typical or green and commercially most valuable Douglas Fir grows extensively throughout western North America, ranging north-south from central British Columbia (lat.55°N) to central Coastal California (lat.36°N), and east-west from 760 to 1,520m (2,500 to 5,000ft) elevation on the western side of the Rockies, the Cascade Range and the Sierra Nevada down to the Pacific Coast, on a wide variety of soils where drainage is good. The trees attain their maximum development in south British Columbia, Vancouver Island, Washington and Oregon, and on some of the western slopes of the Cascade Mountains.

From the northern coastal limits of its range, down towards the centre, Douglas Fir is found growing in mixture with Western Hemlock *Tsuga heterophylla*, Western Red Cedar *Thuya plicata* and Sitka Spruce *Picea sitchensis*. Here the Western Hemlock, being an intense shade-bearer, gradually ousts the light-demanding Douglas until such time as fire sweeps through the forest, after which the Douglas, a fire-climax species, appears again in almost pure stands; at least for a while, but its extreme intolerance of competition soon renders it a sub-climax species, not only with the Hemlocks but also with the Red Cedars, the Spruces and Grand Firs *Abies grandis* – its normal condition in most of its coastal range.

In the more southerly limit of its range Douglas Fir merges with Ponderosa Pine *Pinus ponderosa*, Sugar Pine *P. lambertiana*, Incense Cedar *Calocedrus decurrens*, and in the extreme coastal strip with Redwoods *Sequoia sempervirens*.

In the shrub layer beneath these immense trees a variety of shrubs abound including Vine Maples *Acer circinatum*, Oregon Grape *Mahonia aquifolium* as well as *Rubus* spp. and the invasive Salal *Gultheria shallon*.

Life history

Douglas Fir will start producing cones at about 15 years of age in ideal conditions, and will go on doing so in increasing amounts for up to 300 years. Natural stands, following seed fall, start with more than 2,470 trees to the hectare (1,000 trees per acre) amongst which the dominants will maintain an average annual height growth of more than 60cm (2ft) up to the age of 20, this growth rate gradually tailing off to about 10cm (4in) a year at the age of 120 years.

Although Douglas Fir forest as such is not the tallest forest (this is the achievement of the Coastal Redwoods) it seems possible that the tallest-ever trees were Douglas Firs, as some certainly exceeded 120m (400ft) in height. Evidence of this comes from Walter Draycott's book, *Early Days in the Lynn Valley*, in which he reports being told by an old logger that a Douglas Fir, felled in 1902 on what is now Argyle Road, off Mountain Highway, North Vancouver, measured 126m (415ft) in height and had a diameter at breast height of 4.2m (14ft 3in). The bark was 33.5cm (13.5in) thick and the timber volume was something in excess of 365cu. metres (13,000cu.ft). It seems that the Lynn Valley, Vancouver, was a place of giant trees with an extraordinarily favoured climate and unusually sheltered from the wind, especially oceanic and continental gales, and no doubt there were many other 120m (400ft) Douglas Firs there before the European loggers came upon the scene.

The Douglas Fir is very long-lived, 500 years being not uncommon and 1,000 years not improbable spans of life.

Timber

The timber of Douglas Fir, often called Oregon pine in the trade, is one of the best in the world for a wide variety of uses. It is light, strong, firm and works well. In colour it varies from yellowish to light red, with a narrow band of white sapwood. The massive boles of these trees enable knot-free timber of great sizes to be cut from them, and plywood veneers without blemish, but marked with the typical zig-zag figure, can be peeled from them in huge quantities.

The timber is used for construction work, railway sleepers, house building, etc. and, as the annual rings are reddish and very prominent, the plywood can be attractively stained for many indoor uses.

Introduction, performance and use as an exotic

Douglas Fir, after its discovery by Archibald Menzies in 1793 and its introduction to Britain by David Douglas in 1827, was planted mainly in that country but also in parts of Western Europe, as an ornamental for which its rapid growth and graceful habit made it eminently suitable. But in Britain especially, where strong winds are frequent, they suffered damage to their tops and thus to their appearance as they increased in size. Although there are throughout the country, particularly in sheltered sites, large trees which add greatly to the amenity of the parks and gardens in which they are planted, Douglas Fir is now not much grown as an ornamental tree. And many old specimens which became occluded in expanded urban areas have disappeared because, of all conifers, Douglas Fir is perhaps the most intolerant of air pollution.

Nevertheless, there are up and down the kingdom, and especially in Scotland, a great many very large avenues and others in considerable stands. A tree at the Hermitage, Dunkeld, Perthshire, only 90 years old, and thus probably a natural seedling from the older trees nearby, was 57m (187ft) tall in 1977, and at one time looked like becoming the first tree in Britain to reach a height of 60m (200ft); but when Mitchell measured it again in 1981 it had added only 30cm (1ft) to its stature, although its girth was 3.9m (13ft 2in). It has been overtaken by several trees including one at Moniac Glen, Inverness, which measured 60m × 2.9m (200ft × 9ft 10in) in 1988. Other recently measured trees include the following:-

Belladrum, Inverness, 51.8m × 5.8m (170ft × 19ft 9in) 1987
Blair Castle, Perth, 54m × 4.2m (177ft × 14ft) 1987
Dunans, Argyll, 60m × 5.3m (200ft × 17ft 9in) 1987
Endsleigh, Devon, 42m × 6.4m (138ft × 21ft) 1987
Glenferness, Nairn, 55.5m × 4.5m (182ft × 15ft 4in) 1987
Sluie, Darnaway, 60m × 4.4m (200ft × 14ft 6in) 1987
Tregrehan, Cornwall, 38m × 5.7m (125ft × 19ft 1in) 1987

As a forest tree Douglas Fir is one of the most important exotics grown in Britain. It was one of the first to be introduced from north west America but its early popularity waned as its defects became known. Of these the most serious being susceptibility to wind damage, its decided preference for good soils, if tolerable production was to be maintained

Douglas Fir, Scone Palace, Perth (p.126)

Bhutan Pine, Simla, India (p.130)

Weymouth Pine, Usable Gorge, New York State (p.132)

Monterey Pine, Tregothnan, Cornwall (p.134)

Western Yellow Pine, Yosemite National Park, Calif. (p.137)

Jeffrey Pine, Siskiyou Mts, Calif. (p.139)

(and good soils were at a premium and mainly occupied by well-tried native species), and its liability to late frost damage.

As a long-lived tree reaching great sizes, Douglas Fir needs a good roothold and it therefore follows that it requires for its maximum development a deep, well-drained, porous soil and, incidentally, one which is not too base rich. In general the tree seems to do better in the western and more humid parts of Britain. However, fine trees are to be found in certain favoured areas in the east. Sites on the lower slopes of valleys rather than in valleys themselves (but not necessarily at low elevations), away from late frosts and bad drainage, seem to be most satisfactory, at least for plantings where this species is to be the final crop.

Although Douglas Fir in the natural forests of north west America shows great intolerance to competition from neighbouring, quick-growing giants, it is, in Britain, moderately shade-tolerant vis-a-vis many other forest trees and, provided overhead shade is not too intense, it will put up with considerable side shade. It can therefore be used for restocking old scrub sites.

Some measure of the value of this tree as a timber producer in Britain can be gauged by the fact that on good sites young trees put on 90 to 120cm (3 to 4ft) height growth in a year and maintain this rapid development for several years. Many trees have exceeded 30m (100ft) in 30 years and large numbers have reached 45m (150ft) at the age of 70, although this height growth is not matched by the proportionate increase in girth found in some species. Douglas Fir produces ample seed in Britain and on many sites natural regeneration has been used for restocking, although in general planting is the normal method used.

In Europe, Douglas Fir is generally a healthy tree. Some disfiguring but little lasting damage is caused by fungal diseases and resulting needle-cast, and a sap-sucking, white woolly aphid may cause discoloration and distortion of the needles. Indeed, manifestations of these diseases are probably due to an incorrect choice of site for these trees.

The timber of Douglas Fir grown in Europe seems to be less strong than that from stands across the Atlantic. But like so many exotic conifers which have been introduced in the last 100 years or so, large sawmill sizes are not yet available for fair comparisons to be made. Nevertheless, the home-grown timber has many uses and the largest sizes fetch increasingly high prices.

In the USA the native Douglas Fir is rated as one of the best evergreen ornamental trees, with the inland blue Rocky Mountains form being the hardiest in the north eastern states.

In New Zealand the Douglas Fir has been grown very successfully, in North Island in particular. A 70-year-old stand at Rotorua has a top height in excess of 60m (200ft), and many of the older properties, especially in the Canterbury area, have 100-year-old Douglas Firs in their grounds, the top heights being checked only by the strong winds prevalent in those parts. These trees are also planted in shelterbelts on both islands.

Bhutan Pine

Pinus wallichiana

The Bhutan Pine, from the Himalayas, is one of the most elegantly beautiful of the genus. The 12 to 20cm (5 to 8in) long needles, in bundles of five, are slender, soft and pendulous, hanging from the wide-spreading and drooping branches in bluish-grey masses to give a most pleasing effect to the open, conic crown. The bole, clad in greyish-brown bark, slightly tinged with orange and roughened rather like an oak, is also very distinctive. (See illustration facing p.128)

Distribution and habitat

The natural range of the Bhutan Pine extends right along the Himalayas from Afghanistan in the west to Bhutan in the east, at altitudes varying between 2,440 and 3,800m (8,000 and 12,500ft). At the lower levels in the western part of the range it meets with Chir Pine *Pinus roxburghii* and evergreen oaks *Quercus incana* and *Q. dilatata*; and higher up where it becomes more numerous, some Deodar *Cedrus deodara*, Himalayan Spruce *Picea smithiana* and Pindrow Fir *Abies pindrow* come in. But for the most part these three species, and others, are unable to compete with such a prolific seed producer as the Bhutan Pine, which often manages to form almost pure stands, especially in areas which have previously been ravaged by fire. Higher up still, around the 2,750m (9,000ft) mark and above, in the drier areas not reached by the monsoon, the Bhutan Pine is found growing with Chilgoza Pine *P. gerardiana*, Himalayan Fir *Abies spectabilis*, Sikkim Larch *Larix griffithiana*, Himalayan Hemlock *Tsuga dumosa* and Himalayan Birch *Betula utilis*.

Life history

The female flowers appear two and three together near the tips of the new shoots, and during the summer become small, club-like cones on stalks. During the second summer they grow pendulous and elongate into 25cm (10in), cylindrical, light brown cones, opening their thick scales to shed their seeds.

Seedling growth is very rapid, more so than those of competitors, averaging 60cm (2ft) a year for the first 60 years or so. Occasionally mature trees may be 45m (150ft) tall and 3.6m (12ft) in girth, but more usually 30 to 36m (100 to 120ft) tall and girthing 2 to 3m (6 to 10ft).

This pine is an important species in the managed forest in the west of the range, both in mixture with other trees and in pure stands managed under the uniform system.

Timber

Next to that of the Deodar, the timber of the Bhutan Pine is the most highly valued for house building, railway sleepers and other local uses; and as it contains less oil, it is used for tea and apple boxes.

Introduction, performance and use as an exotic

The Bhutan Pine was introduced into Britain by Lambert who raised many plants from seed on his estate at Boyton, Wilts. in 1823. Wallich sent home more seed in 1827 from which plants were raised at the Horticultural Society's gardens at Chiswick and at the Glasgow Botanic Gardens.

Loudon mentions a Bhutan Pine growing in the Chiswick Gardens which was 3.6m (12ft) tall and eight years old in 1837; and another at Dropmore, Bucks. which was 3m (10ft) tall in the same year. However, this does not seem to be the same tree as the one at Dropmore which Mitchell measured there in 1970 as being 27.4m × 2.4m (90ft × 8ft), for he gives the planting date as 1843. But the best specimen which Elwes found was at Hewell Grange, Worcs. which, in 1909, was 28.3m × 2.4m (93ft × 8ft 4in). It is the same tree as that measured by Mitchell in 1963 as 27.4m × 3.2m (90ft × 10ft 8in) and it has continued to go back in the crown, although increasing in girth, for in 1974 it measured 26m × 3.3m (85ft × 11ft 1in). It was planted in 1831 and is thought to be one of the originals. No other is known today, but there are some fairly old specimens still alive, which Elwes measured, such as the one at The Frythe, Welwyn, Herts., which, in 1906, was 18.3m × 2m (60ft × 7ft) with very widespread branches, some of them layered to give a total crown circumference of 75m (246ft). In 1969 this tree measured 21.3m × 2.4m (70ft × 8ft 2in); and one at Wilton House, Wilts., which was 23.4m × 2.4m (77ft × 8ft 3in) in 1906, had grown to 29.5m × 3.2m (97ft × 10ft 6in) by 1971.

There are some fine trees of later plantings such as that at Bicton, Devon, which in 1968, measured 34m × 4.7m (112ft × 15ft 9in); and at Redleaf, Kent measuring 32m × 2.6m (105ft × 8ft 10in) in 1963. Some very young trees, particularly those at Albury, Surrey, have grown very quickly, reaching 28.3m and 24m (93ft and 79ft) in height respectively in 47 years.

A selection of the most recent measurements include the following trees:-

Abbeyliex, Co. Leix, 37m × 3.4m (121ft × 11ft 8in) 1987
Albury, Surrey, 31m × 2.2m (102ft × 7ft 8in) 1987
Fota, Co. Cork, 31m × 4.2m (102ft × 14ft 3in) (at 90cm (3ft)) 1987
Longnor, Salop, 24.4m × 3.2m (80ft × 10ft 9in) 1987
Melbury, Dorset, 25m × 3.3m (82ft × 11ft) 1987
Stowe, Bucks., 35m × 3m (115ft × 10ft 1in) 1987
Wilton House, Wilts., 21.3m × 3.2m (70ft × 10ft 10in) 1987

The Bhutan Pine is the most successful of the Asiatic pines grown in Britain. In youth and middle age it is a very desirable ornamental with its long, bluish needles elegantly draped on pendulous branches; but old trees tend to die back and lose their attraction.

It was introduced into the USA in 1827, but in many areas it seems to become damaged by winter cold and strong winds. However, it does particularly well in Philadelphia and Seattle, and flourishes in other places where the growing conditions are suitable.

This pine is less affected by air pollution than many conifers.

Weymouth or White Pine

Pinus strobus

The Weymouth Pine is the largest conifer growing in eastern North America and, after the Sugar Pine, the largest pine in the whole of North America. The beautifully straight, gradually tapering trunk, enclosed in thick, rough, dark grey bark, leads up to a broad, irregular crown of thin branches, horizontal at first but sweeping upwards at the tips. The short, blue-green needles in bundles of five are grouped towards the ends of the branches, at heights of up to 45m (150ft). (See illustration facing p.128)

Distribution and habitat

The natural range of the Weymouth Pine stretches across southern Canada from Manitoba to Newfoundland and through the north and eastern USA from Minnesota and north eastern Iowa to the Atlantic Coast; and south along the Appalachian Mountains to North Georgia and Alabama. It grows on most soil types throughout the range but fares best on well-drained sandy loam along rivers and streams. In the north it is found between sea level and 460m (1,500ft), and in the southern Appalachians on northerly aspects in a band between 360 and 1,060m (1,200 and 3,500ft), and occasionally even at 1,220m (4,000ft) in sheltered valleys.

As this pine is perhaps less gregarious than others, pure stands usually result from fire or man-made clearing; it is more often found as a natural component of mixed conifer and deciduous forest. In the north east, on sandy soils, it consorts with such species as White Birch *Betula papyrifera*, Balsam Fir *Abies balsamea*, Large-toothed Aspen *Populus grandidentata* and Red Spruce *Picea rubens*. Further south, where it reaches its maximum development, it consorts with black and white oaks, Eastern Hemlock *Tsuga occidentalis* and Sugar Maple *Acer saccharum*.

Life history

The slim, pendant cones are borne at the early age of 10 years, the production becoming very prolific at about 50 years when a single tree may carry as many as 400 cones in one year. The seeds are shed in the autumn and good seed years occur every three to five years; but a cone weevil has caused great damage in some areas, preventing seed production for as many as 10 consecutive years.

Seedling growth is slow at first – 12cm (5in) in three years, and 1.2m (4ft) in 10 years being normal, but after the first decade growth speeds up and may be maintained at as much as 1.2m (4ft) a year for some time.

Weymouth Pines can reach great sizes. Early loggers reported trees as tall as 75m (250ft) and stumps of 2m (7ft) in diameter, but nowadays none of these huge trees, which were probably at least 450 years old, is to be found; even trees 45m (150ft) tall are not common.

This pine is reasonably shade tolerant and, in mixture with other pines and birches, may become dominant; but with oaks and maples it is usually shaded out and killed.

Timber

The white to reddish-brown, soft, straight-grained timber is easily worked and is one of the most economically important woods in North America, used for almost all general purposes from ships' masts to matches, and from crates to construction work.

Introduction, performance and use as an exotic

The Weymouth Pine was first cultivated in England by the Duchess of Beaufort at Badminton in 1705. A little later some were planted by Lord Weymouth at Longleat, which prospered and were given his name, at least in Britain. Yet others were planted at Whitton, and these started to bear seed in 1720, the seeds finding their way to nurserymen. Thus the species became popular.

Many of these early specimens were about 24m (80ft) tall when Loudon saw them around 1838, but Elwes, about 70 years later, does not seem to have found any of them or indeed any of the younger ones listed by Loudon.

The tallest which Elwes measured was a specimen at Stowe, Bucks, which was 31.7m × 3.9m (104ft × 13ft 2in) and, he reckoned, 150 years old. Another old tree at Pains Hill, Surrey, also missed by Loudon, was 27.4m × 3.7m (90ft × 12ft 8in) in 1904 and much the same size when measured by Gardner in 1947.

Only three trees mentioned by Elwes had survived until the 1970s: at Ombersley Court, Woburn Park and Nuneham Court. Although all three showed considerable increases in girth in the 60 or so years between recorded measurements, only one increased in height.

In 1969, the tallest Weymouth Pine in Britain, according to Mitchell, was one at Pucks Pit, New Forest, which measured 39m × 2.5m (128ft × 8ft 5in). More recent measurements include a taller tree:-

Blair Drummond, Perths., 26m × 3.3m (85ft × 11ft 5in) 1987
Brookwood Necropolis, Surrey, 25m × 2.5m (82ft × 8ft 6in) 1989
Cannop, Glos., 37m × 3.4m (121ft × 11ft 8in); 37m × 3m (121ft × 10ft 4in) 1988
Chatsworth, Derbys., 44m × 2.6m (144ft × 8ft 10in) 1989
Kitlands, Surrey, 26m × 3.8m (85ft × 12ft 9in) 1987
Longleat, Wilts., 28m × 2.9m (92ft × 9ft 10in) 1986

As a forest tree, in the early years, this pine was tried with considerable success, but the blister rust disease struck and, by 1909, was devastating plantations to such an extent that the authorities advised against further planting. The advice holds good to this day. Barring this disease, the Weymouth Pine would have a great future in Britain, for measurements of the few healthy trees suggests it is likely to attain sizes comparable to those in America in perhaps fewer years. In Continental Europe, where there are considerable plantings, it seems less susceptible to the blister rust.

In the USA this native pine is held in great regard as an ornamental, quite apart from being the fifth most important timber in that country. Wyman writes:-

'White Pine may grow 10 feet (3m) in 10 years and 25 feet (7.6m) in 20 years if it is on good soil. Young trees are often pyramidal in habit but the older the tree grows the more picturesque it becomes – with flat top and a missing branch here and there until at a ripe old age these trees are the most conspicuous (and beautiful) in almost any landscape.'

By careful pruning, the growth of this pine can be restrained and thus it need not grow out of scale even in a small garden.

Monterey Pine

Pinus radiata

The Monterey Pine is a coastal species with a very limited range in California, and conspicuously different from other American pines because of its bright green, slender needles (in bundles of three) forming an open, round-topped crown, usually no more than 18m (60ft) tall. The trunk, when open-grown, has big low-down branches; but when forest grown, is often disfigured with persistent, broken dead branches. Old cones remain in whorls along the branches and even on the trunk, for as long as 30 years. The bark is dark red-brown and deeply fissured. (See illustration facing p.129)

Distribution and habitat

The natural range of the Monterey Pine is confined to three locations on the coast of California. The largest stand of some 4,050 hectares (10,000 acres) is on the Monterey Peninsula, and another of 1,000 hectares (2,500 acres) surrounds the town of Cambria 136 kilometres (85 miles) south of Monterey, with a small area of 200 hectares (500 acres) some 9 kilometres (6 miles) north of this town. The third main area is north of Monterey, some 22 kilometres (14 miles) south of Santa Cruz and covers less than 400 hectares (1,000 acres).

This tree seems to require high humidity and is never found much more than 9 kilometres (6 miles) from the coast, and usually on soils which are sandy and coarse textured, on gentle slopes below the 300m (1,000ft) contour.

This pine is not dominant in the north, but grows associated with Redwoods *Sequoia sempervirens* and Douglas Fir *Pseudotsuga menziesii* on the middle slopes, and on drier sites with Knob-cone Pine *Pinus attenuata*. Further south, on the coast, Monterey Cypress *Cupressus macrocarpa* and Goven's Cypress *C. goveniana* are associates, and inland, where the stands are mainly pure, Bishop Pine *P. muricata* is occasionally found with an understorey of Californian Oak *Quercus agrifolia*.

Life history

The cones are borne in clusters on the main stem and branches, and mature in the autumn of the second season; the seeds often do not fall until the following spring. Although, in some cases, trees as young as four years may produce cones, real production does not start until the trees are aged about 15 years, and then it is prolific – a tree around 15m (50ft) tall may produce over 6,000 cones. The viability of the seed is high, around 50%, and dense carpets of seedlings result. Growth is rapid – up to 30cm (1ft) in the first year, with three-year old trees 1.8m (6ft) tall.

In native stands this pine rarely exceeds 30m (100ft) in height and 2.7m (9ft) in girth; but in New Zealand and Australia 36m (120ft) is usual and 60m (200ft) has been recorded.

Although the natural stands are small and mostly pure, there is reason to think that the Monterey Pine is a shade-tolerant tree producing, in the natural state, dominant individuals standing well above the general canopy level early in life, suppressing competitors of the same species and of others. In New Zealand, plantations thin themselves quite effectively by this natural process.

Timber

In America the fairly light, hard (for pine) and strong timber is in such short supply that it is of no commercial importance. However, that grown in New Zealand, Australia, South Africa and Spain is an excellent general purpose timber used for building, furniture, crates and pulp.

Introduction, performance and use as an exotic

The Monterey Pine was discovered by David Douglas during his stay in Monterey in 1831-2, and he sent seeds home in 1833. Some plants were raised from these by the Duke of Devonshire at Chiswick which were 0.9 to 1.5m (3 to 5ft) tall when Loudon saw them in 1838.

Elwes records a specimen at Dropmore, Bucks. raised from a cutting (presumably from one of these original trees) in 1839, which was 23.4m × 4.5m (77ft × 15ft) in 1909. This same tree was 27.4m × 5m (90ft × 17ft 2in) when measured by Mitchell in 1970.

The tree at Cufnells, New Forest, which Elwes noted as the most remarkable he had seen, and of which he took a photograph in 1907 when it was 35.3m × 2.5m (116ft × 8ft 6in), with a clear bole of 12m (40ft), was 44m × 3.6m (144ft × 12ft 2in) when Mitchell measured it in 1970; but 10 years later it had made no more height growth and its girth had only increased by 12cm (5in). Another specimen photographed and measured by Elwes at Goodwood, Sussex, in 1906 as 25.3m × 2.7m (83ft × 9ft) had grown to 28.3m × 4m (93ft × 13ft 6in) by 1965.

Although these old trees show the Monterey Pine to be long-lived in England, it is clear from measurements made by Mitchell of 18 Elwes and Henry trees, as well as many not seen by them, that they are unlikely to grow to much more than 36 to 39m (120 to 130ft) here. And from Mitchell's measurements of younger trees it appears that the rapid growth occurs in the early years. A tree at Albury, for instance, planted in 1926, reached a height of 24m (79ft) in 28 years.

Some recent measurements include the following trees:-

Bodnant, Gwynedd, 33m × 4.5m (108ft × 15ft 1in) (planted 1900) 1981
Carclew, Cornwall, 26.8m × 5.5m (88ft × 18ft 5in) 1987
Kelburn Castle, Ayr, 29.8m × 5.2m (98ft × 17ft 8in) 1987
Tregothnan, Cornwall, 36.6m × 6.7m (120ft × 22ft 2in) 1987
Woodstock, Co. Kilkenny, 37m × 5.5m (121ft × 18ft 7in) 1987

The best of the ornamental specimens of this pine are found in southern England, especially along the coast from Hampshire westwards, but cold winds cause extensive but apparently harmless browning of the needles.

As a forest tree in Britain it may have a place, by virtue of its quick growth, in the extreme south west of England, but the persistent dead branches on the stems will need careful pruning.

In the USA this native pine is extensively planted as an ornamental in western coastal areas.

In Australia, New Zealand and South Africa it is one of the most successful exotic forest trees; very fast-growing and yielding large volumes of utility timber. In New Zealand it is the most important commercial species, forming 60% of the conifers grown there and covering more than 400,000 ha (1 million acres). In Australia about half a million acres are planted with it, and in South Africa it is a species of major importance.

In these sub-tropical and warm temperate regions the growth rate is almost unbelievable. Top heights of 36m (120ft) and timber volumes of 7,600cu. metres per hectare (11,000cu.ft per acre) in 30 years are common achievements.

Western Yellow or Ponderosa Pine

Pinus ponderosa

The Ponderosa Pine is one of the largest and most widely distributed pines in North America. Its immense trunk is covered in orange-yellow bark split into huge, irregular plates. A spire-like crown, up to 60m (200ft) tall, is formed by the short thick, often pendulous branches, which bear the densely crowded long yellow-green needles in bundles of three. (See illustration facing p.129)

Distribution and habitat

The natural range of this pine covers a tremendous area from as far north as the Fraser River in British Columbia, right down to Durango, Mexico, in the south, and between these two points down both sides of the Pacific and Rocky Mountains at elevations varying from sea level in Washington up to 2,750m (9,000ft) in California and Colorado. It grows on a variety of soils but generally develops best on well-drained, deep sandy gravels.

With such a wide range, Ponderosa Pine tends to be a component of many different forest types, as well as being found in pure stands in the more southerly parts. In the Pacific and westerly part of the range it grows among Douglas Fir *Pseudotsuga menziesii*, Western Larch *Larix occidentalis*, and Grand Fir *Abies grandis*; whilst further south and inland such trees as Sugar Pine *P. lambertiana*, Incense Cedar *Calocedrus decurrens* and Blue Spruce *Picea pungens* grow with it. On some of the highest areas Bristle-cone Pine *P. aristata* may accompany it.

Life history

The cones, 7 to 15cm (3 to 6in) long, are borne on trees from 16 years old upwards, those between 60 and 100 being the most productive, with trees up to 350 years old still producing seed. Heavy crops occur every eight years or so, but a great many seeds are eaten by birds, squirrels, mice and chipmunks.

Seedlings have vigorously growing tap roots – 55cm (22in) in one year – and a top height of 5cm (2in). After four years the root may be 150cm

(60in) long and the seedling 30cm (12in) tall, all of which gives it a great advantage in dry areas.

The Ponderosa Pine grows to an impressive size.Heights of 27 to 36m (90 to 120ft) and girths of 2.7 to 3.6m (9 to 12ft) are usual, and trees of 70m (230ft) and 7.3m (24ft) in girth have been recorded. It is also long-lived: old stands often contain trees of between 300 and 600 years of age. One specimen in Oregon was accurately measured as being 726 years old.

Timber

The timber is hard, strong, fine-grained and light red in colour. Being available in huge quantities, it has many uses including general construction, interior finish, poles, boxing, crates and pattern making.

Introduction, performance and use as an exotic

This pine was first seen by Lewis and Clarke who found it in Upper Missouri during their journey across the Rockies in 1804. But it was not described until David Douglas had found it on the Spokane River, Washington, in 1826. He introduced it into Britain the following year and a number of plants was raised from the seed he sent home. One of them, at Dropmore, Bucks., was mentioned by Loudon as being 2.7m (9ft) tall in 1837, and Elwes says that it was 30m × 2.6m (99ft × 8ft 9in) in 1909. It had gone back a foot or so by 1970 when Mitchell found it to be 29.8m × 2.9m (98ft × 9ft 10in).

There were three other trees from the original batch still living in 1970: at Arley Castle, Bowood and Powis Castle, all planted in 1829. The Powis tree was, in 1909, the tallest known to Elwes, and still is the tallest in Britain, being 41m × 4.1m (135ft × 13ft 9in) when measured in 1981. The Bowood tree was also living in 1989 when it measured 41m × 4.3m (135ft × 14ft 5in). A specimen at Bayfordbury, Herts. which Elwes found to be 30m × 2.7m (100ft × 9ft 4in) in 1906, was blown down in 1951 when it measured 37.8m × 3.3m (124ft × 11ft).

In 1970, Mitchell gave details of some 40 other trees, between 50 and 120 years of age, and mostly in southern and central England and very few in the west. All seem to have grown about a foot a year. More recent measurements include the following trees:-

Albury Park, Surrey, 32m × 3m (105ft × 10ft 4in) 1987
Bodnant, Gwynedd, 22m × 2.7m (72ft × 9ft 3in) (planted 1901) 1987
Dawyck, Peebles, 35m × 3.2m (115ft × 10ft 8in) 1988
Eastnor Castle, Herefords., 36m × 3.4m (118ft × 11ft 5in) 1987
Powis Castle, 37.8m × 3.2m (124ft × 10ft 9in); 33.8m × 4.6m (111ft × 15ft 6in); 40.3m × 4.2m (132ft × 14ft 3in) 1988

This pine is not often planted as a forest tree in Britain, as it seems to favour sites suitable to Corsican Pine, but does not really match up to the performance of this well-tried European species.

As an ornamental, Ponderosa Pine is chiefly remarkable for its attractive yellowish and heavily plated bark. The tree is fairly common in some of the collections and larger gardens in the south of the country. Loudon mentions that a tree at Dropmore and another of the originals at Chiswick had, during the first 10 years of their lives, suffered from attacks of beetles, but this does not seem to have become a chronic affliction, for most of the old trees which have been growing in England for 150 years and more are perfectly healthy.

In the USA this native pine is planted as an ornamental because of its quick growth and its yellowish-green foliage. A reputed variety *scopulorum*, which is found in the Rocky Mountains, is smaller in all parts and is hardier and more columnar in habit, according to Wyman. But these characteristics may only be the result of environment and adaptation. At all events, it is more suitable for confined spaces in the colder areas.

Jeffrey Pine

Pinus jeffreyi

The Jeffrey Pine, a tree of the mountains of North West America, is in many respects very like Ponderosa Pine, with its needles in bundles of three, but bluish-green instead of yellow-green and rather stiffer. It has the same conic and shapely crown, but usually with smaller branches. The main difference is in the bole which has very dark red, almost black, bark; the cones are also much bigger. (See illustration facing p.129)

Distribution and habitat

The natural range of the Jeffrey Pine extends from southern Oregon in the north, southwards along both slopes of the Coast Range and the Sierra Nevada to southern California, just touching western Nevada.

Although this range overlaps almost completely that of the Ponderosa Pine, Jeffrey Pine is found growing pure in frost pockets and on high, cold ridges where its relative cannot prosper. Only in the north does it grow below 300m (1,000ft), and the best stands are found between 1,520 and 1,980m (5,000 and 6,500ft) with an occasional tree at 3,050m (10,000ft).

Common associates, apart from *P. ponderosa*, are Douglas Fir *Pseudotsuga menziesii*, Red Fir *Abies magnifica*, White Fir *A. concolor*, Incense Cedar *Calocedrus decurrens* and Sugar Pine *Pinus lambertiana*; with Bigcone Pine *P. coulteri* and Californian Live Oak *Quercus agrifolia* in the extreme south.

Life history

The big cones, 12 to 37cm (5 to 15in) long, are much bigger than those of the Ponderosa Pine, and ripen in the September of the second year. The seeds, which are twice the size of those of its near relative, are released in October and November, and have a very high germination rate. The seedlings grow quickly, reaching a height of 1.2 to 1.5m (4 to 5ft) in 10 years, speeding up in the second five years.

Although this pine habitually looks more vigorous and retains a youthful appearance longer than the Ponderosa, its growth rates are about the same, the Jeffrey reaching a maximum height of some 60m (200ft) and a girth of about 4.5m (15ft), and with a long life span, 400 to 500 years being quite common. A 600-year-old tree is known to have been felled.

Like Ponderosa, Jeffrey Pine is tolerant of shade, but it is often found in pure stands where climatic conditions favour it, or in mixture with Ponderosa and other species. When competing with the more shade-tolerant White Fir in high rainfall areas, it loses any dominance it might have elsewhere.

Timber

Although the timber of Jeffrey Pine tends to be lighter and softer than that of Ponderosa, it is marketed for the same purposes.

Introduction, performance and use as an exotic

The Jeffrey Pine was discovered by John Jeffrey in 1852 and it was introduced into Britain in the same year by the Oregon Association.

At the turn of the century, when Elwes was writing, the Jeffrey Pine was considered to be a variety of Ponderosa, and there was obviously some confusion over naming specimens. A tree at Scone Palace, Perthshire, mentioned by Elwes, measured 15.2m × 1.9m (50ft × 6ft 8in) in 1891. It was not specified as var. *jeffreyi*, but may be the same one that Mitchell gives as the finest Jeffrey Pine in Britain. It was planted in 1860 and measured 36m × 3.6m (118ft × 12ft 5in) in 1970. By 1989 it had grown to 39.6m × 4.1m (130ft × 13ft 10in) and retains its paramountcy. None of the var. *jeffreyi* specimens mentioned by Elwes seem to have survived, and Mitchell reports that this pine does not seem to be long-lived, many of the old trees clearly being on the way out. He suggests that four trees on two adjacent Surrey Estates, all of which exceed 3m (10ft) in girth and are between 21 and 24m (70 and 80ft) tall, may well be from the original seed. Other notable trees include the following:-

Adhurst St Mary, Hants., 35m × 2.6m (115ft × 8ft 10in) 1987
Eastnor Castle, Herefords., 29.8m × 3m (98ft × 10ft 4in) 1987
Peper Harow, Surrey, 29.8m × 3.6m (98ft × 12ft) 1987
Rossie Priory, Perths., 37m × 2.7m (121ft × 9ft 3in) 1988

As an ornamental in Britain today it is widespread but not common, being confined mainly to large gardens and collections.

In the USA this native pine is planted as an ornamental only on the west coast.

Austrian pine

Pinus nigra var. *nigra*

The Austrian Pine is a similar sized tree to the Corsican Pine but with a more domed and heavily branched crown, bearing needles which are much more bunched and darker of hue, and often separated by lengths of bare branches. The bark too is much darker, being brown to blackish and split by flaking fissures into scaly plates. (See illustration facing p.144)

Distribution and habitat

This pine occurs naturally from central Italy north east to lower Austria and western Hungary; south through Yugoslavia and Albania to Greece, as well as eastwards into western Bulgaria. It is for the most part an alpine tree, the best stands growing at around 920m (3,000ft), but also as high up as 1,620m (5,300ft). At lower elevations it consorts with Beech *Fagus sylvatica*, albeit under threat in some circumstances as I described in *Quarterly Journal of Forestry* (67:1):-

> 'In the Tara Valley (in Yugoslavia) we made a short stop in a memorable and ancient stand of Black Pine *P. nigra*. These 400-year-old trees, some standing at 42m (137.8ft) and all of considerable girth and volume, were in a state of siege by beech forest. Four centuries ago they had appeared as a fire-climax species in an area of devastation and had prospered in the midst of their insidious broadleaved brethren by virtue of their quick growth in early years. But no regeneration had been possible because of the everlapping tide of beech encroachment beneath their canopy. Now they stand with their handsome boles engulfed in a dense understorey of beech, doomed to eventual and not too distant extinction. We were glad to have seen them, for they must be some of the finest Black Pine in Europe.'

Between 920 and 1,220m (3,000 and 4,000ft) the Austrian Pine forms almost pure stands with thick natural regeneration, but higher up, particularly on the western slopes where the rainfall is greater, it gives way to Norway Spruce *Picea abies* and Silver fir *Abies alba*. On limestone formations in parts of Yugoslavia it is accompanied by Serbian Spruce *P.*

omorika. It also forms a minor component in mixed forests at lower altitudes on better soils, as I reported in the same *Journal*:-

> 'Once again our drive, this time on forest roads, took us through a fascinating horizontal range of trees. On lower slopes Sessile and Turkey Oaks with Beech, Hornbeam *Carpinus betulus* and Ash, gradually gave way to Norway Spruce, Black Pine, Fir, Birch and Aspen *Populus tremula*, with Scots Pine coming in on the higher slopes, and with Beech very often present in some quantity in the more sheltered cwms. The proportion of the species in these lovely mountain forests was: Spruce 33%, Black Pine 34%, Silver Fir 26% and Beech 6%.'

Life history

There is little difference between the early growth rate of Austrian and Corsican Pine, nor is there any marked difference between their ultimate sizes. Elwes saw trees in Lower Austria which were 30m (100ft) tall and said to be 280 years old. In Yugoslavia he found trees in excess of 36m (120ft) tall; and, as referred to above, I photographed a grove of Black Pines in the Tara Valley which were nearly 42m (140ft) tall, albeit drawn up by competing Beech, but of considerable girth none the less. As to age, there was a tree in Lower Austria which had a ring count of 584, although it was only 20m (65ft) tall and 1.8m (6ft) in girth.

Introduction, performance and use as an exotic

The Austrian Pine was introduced into Britain in 1835 by Lawson of Edinburgh, more than 75 years later than the Corsican, which is surprising. Loudon, writing a mere three years after the introduction, could not record any specimens, but he mentions that seeds sown by Lawson on sandy soil produced seedlings twice as large as those of Scots Pine sown at the same time.

Elwes mentions four trees, three of which Mitchell was unable to trace, and the fourth which he did, turned out be a Corsican. He also writes of two trees, planted at Dawyck in 1840, which may well be originals from Lawson's seed. In 1961, they measured 29m × 2.8m (95ft × 9ft 6in) and 22.8m × 2.4m (75ft × 8ft) respectively.

The tallest Austrian Pine which Mitchell listed in 1970, was one of unknown planting date at Bolderwood in the New Forest which measured 35m × 3.7m (115ft × 12ft 7in). Ten years later it had added 90cm (3ft) to its height. Other recent measurements include the following trees:-

Brocklesby Park, Lincs., 32m × 4m (105ft × 13ft 8in) (1977)
Dawyck, Peebles, 37.8m × 3.1m (124ft × 10ft 5in) (1975)
Glasnevin Botanic Gardens, Dublin, (planted in 1888), 25m × 3.1m (82ft × 10ft 6in) (1980)
Gwysany, Clwyd, 26.8m × 3.7m (88ft × 12ft 6in) (1976)

Harpton Rectory, Powys, 29m × 3.9m (95ft × 13ft 4in) (1978)
Mountfield Court, Kent, 33.5m × 4.2m (110ft × 14ft 3in) (1981)
Oxenford Grange, Surrey, 24.4m × 4.2m (80ft × 13ft 10in) (1978)
Oxford Botanic Gardens, 23.4m × 3.9m (77ft × 13ft 3in) (1978)
Sidbury Manor, Devon, 24.4m × 3.7m (80ft × 12ft 5in) (1977)
Westonbirt, Glos., 31m × 3.2m (102ft × 10ft 7in) (1980)

According to Wyman, the Austrian Pine was introduced into the USA in 1759, but the date suggests that it may well have been the Corsican. It is used for ornamental planting, and a cultivar 'Pyramidalis' is popular there. In some areas it seeds itself and has spread naturally.

Corsican Pine

Pinus nigra var. *maritima*

This variety of the Black Pine is a large tree with a narrowly conic crown of long, grey-green needles on level, evenly spaced branches (except in old trees when the lower branches tend to droop with their weight) supported on a straight bole with deeply fissured, dark grey bark. (See illustration facing p.144)

Distribution and habitat

The Corsican Pine occurs naturally in three main somewhat mountainous areas: in Corsica, Calabria (the southern 'toe' of Italy) and in Sicily where, on mainly poor, dry soils of granite debris origin, it forms extensive forests at altitudes between 820 and 1,830m (2,700 and 6,000ft). At the lower elevations the pine forests are often contiguous with dense woods of Holm Oak *Quercus ilex* and with scattered groves of Downy Oak *Q. pubescens*. In some areas this pine mingles with Maritime Pine *P. pinaster* which occupies the lower and coastal zones. Over most of the forest Corsican Pine grows pure except in ravines where the soil is less poor and moister when some Silver Fir *Abies alba* and Beech *Fagus sylvatica* may occur along with Alder *Alnus cordata*.

Life history

The female cones appear in June in twos and threes borne on very short stalks on the tips of the young branchlets, but they do not ripen until the end of their second year, then they cast their winged seeds in the spring and summer of their third year. First year seedlings seem remarkably small and insignificant, but after two or three years manage to develop shoots up to 90cm (3ft) long. Steady growth continues thereafter in prime conditions. In Calabria, towards the southernmost part of the range, trees 80 to 90 years old can be up to 27m (90ft) tall, with girths of 2 to 2.5m (6 to 8ft). But further north, in Corsica, and at higher altitudes where the pines grow in pure forests, growth is slower, although large sizes can be attained in sheltered and more favourable situations. Elwes mentions a tree growing at 980m (3,200ft) which measured 44.2m × 3.6m (145ft × 12ft 3in); and another with a short bole which girthed 7m (23ft). These trees would probably be very slow grown and aged between 350 and 700 years, the latter being about the maximum lifespan for this pine.

Timber

The timber is similar to that of Scots Pine but somewhat coarser with a larger proportion of sapwood and inclined to have larger knots with consequently less resistance to shock impacts. Its main uses are for estate work after preservation treatment, although it is fairly resistant to this, and for mining timber and fibreboard manufacture.

Introduction, performance and use as an exotic

According to Loudon, the Corsican Pine was introduced into Britain in 1759 under the name of *Pinus sylvestris altissima*, a name adopted by Alton in the first edition of *Hortis Kewensis*. Seeds were not readily obtainable until the 1820's when a general distribution throughout the country was made. In consequence, few of the trees mentioned by Loudon in his *Arboretum et Fruticetum Britannicum* of 1834 were much more than a dozen years old and thus of no great size, except for one at Kew which was 24m (80ft) tall and must have been one of the 18th-century introductions. The 6m (20ft) specimen at Dropmore, which Loudon mentions could be one of those recorded by Mitchell as being 36.6m × 4.2m (120ft × 14ft) in 1970.

Elwes records a tree at Kew, planted in 1814 and considered to be the oldest in England. It measured 26.2m × 2.7m (86ft × 9ft 3in) in 1903, and 27m × 2.8m (89ft × 9ft 6in) when Mitchell measured it in 1970. By 1981 it had grown to 28m × 3m (92ft × 9ft 10in).

Elwes makes no mention of Loudon's 18th century tree which presumably had died. Of six fine specimens at Arley Castle, planted in 1820, all over 30m (100ft) tall in 1903 (one of which was photographed by Elwes) three were still there in 1961, the tallest being 38m × 4.2m (125ft × 14ft 3in).

Some of the earliest of these pines, planted in the first decade of the 19th century, were at Holkham Hall, Norfolk. Three of these are now over 30m (100ft) tall and seedlings from them were used to plant up sand dunes on the nearby Norfolk coast. These plantations are now extensive and self-perpetuating.

Recent measurements include the following trees:-

Aigas House, E. Ross, (planted in 1877), 36m × 2.7m (118ft × 9ft) 1980

Austrian Pine, Troodos Mts, Cyprus (p.141)

Japanese Black Pine, Wootton Lodge, Staffs. (p.145)

Corsican Pine, nr. Harwell, Berks. (p.143)

Lodgepole Pine, Bedgebury Pinetum, Kent (p.147)

Lombardy Poplar, Newly Hall, Yorks (p.149)

Black Cottonwood by riverside, Pemberton, British Colombia (p.151)

Benenden School, Kent, 33m × 3.7m (108ft × 12ft 7in) 1979
Bicton, Devon, 33.8m × 3.7m (111ft × 12ft 7in) 1977
Blenheim Palace, Oxon., 35m × 3.4m (115ft × 11ft 8in) 1978
Brockett Hall, Herts., 37m × 3.4m (121ft × 11ft 5in) 1976
Cuffnels, Hants., (planted in 1856), 39.6m × 3m (130ft × 10ft 3in) 1980
Holkham, Norfolk, 32m × 4.2m (105ft × 14ft) 1973
Leaton Knolls, Shrops., 40.2m × 3.3m (132ft × 10ft 10in) 1981
Skibo Castle, Sutherland, 38m × 2.1m (125ft × 7ft 4in) 1980

In the mid-19th century, when Larch had failed largely because of canker in badly chosen provenances, Corsican Pine became popular as a plantation tree, only to be superseded by Douglas Fir, especially on better soils. However, the pine's tolerance of poor soil conditions, particularly in sandy coastal areas, and its acceptance of calcareous soils, have resulted in extensive plantings, especially in southern and eastern counties; in Thetford Forest, for instance, where it responds to the warmer, drier summers.

This variety of Black Pine does not seem to be much planted in the USA; certainly not as much as the Austrian variety.

Japanese Black Pine

Pinus thunbergii

The Japanese Black Pine is a large tree, usually with a somewhat crooked trunk. It supports horizontally spreading branches which are also often crooked. The stout, yellowish-green needles, borne in pairs are densely arranged on the branches and surround uniquely white, silky buds. The dull grey bark is deeply fissured. (See illustration facing p.144)

Distribution and habitat

This pine is essentially a maritime species and indigenous only to Japan. Although originally growing wild only on the east coast from Kyushu northwards to central Honshu, it has become, by planting over the centuries, one of the commonest trees of the central and southern parts of the country. In the north it is mainly found in coastal districts, but in Shikoku and Kyushu it grows well inland, often in association with the Japanese Red Pine *P. densiflora*, and natural hybrids are common.

The Black Pine is extensively planted in Japan for stabilising sand dunes and checking erosion on the lower mountain slopes near the coast. It is also, because of its irregular shape, a popular subject in Japanese gardens.

Life history

The flowers appear in April, the females developing into clusters of up to 50 cones, purplish and quite small at first, but becoming brown and woody and some 6cm (2.5in) long when they ripen and shed their winged seeds in the second autumn.

Early growth is rapid – up to 45cm (18in) a year – but later tails off. Even so, some trees reach enormous dimensions, topping 43m (140ft) with girths of 3.6m (12ft). Elwes describes a celebrated tree in the Kyoto area which had a girth at ground level of 6m (20ft) and, at 1.5m (5ft) up, 8.8m (29ft) where a large limb emerged. This branch and others were themselves 3.6m (12ft) in girth, with 6m (20ft) of trunk above the uppermost, and the whole crown had an enormous spread 54m (180ft) in circumference. This tree was said to be 1,000 years old.

It is this Black Pine which is so commonly depicted by Japanese artists.

Timber

The timber, although less easily worked than that of the Red Pine, is more resinous and durable. It is used for construction work, railway sleepers and for the foundations of wooden buildings. The trees are also tapped for resin.

Introduction, performance and use as an exotic

The Japanese Black Pine was first introduced into Europe by Siebold who brought it to Holland in 1855. J. Gould Veitch brought it to Britain in 1861.

Elwes noted, at the turn of the century, that it was often to be seen in private collections and botanic gardens, but none of the five trees which he listed is now alive. The best of them was a specimen at Dropmore, Bucks., planted in 1861, which measured 15.8m × 1.8m (52ft × 6ft) in 1909. He also reported that these pines had been found to be superior to Stone Pine *P. pinea* for coastal planting in Guernsey where they had grown to 7.6m (25ft) in 20 years.

In Britain generally this pine does not seem to be very hardy or windfirm, although specimens are to be found in many large collections and gardens. The oldest, according to Mitchell, was a tree at Westonbirt, Glos., which was planted in 1876 and measured 18.6m × 1.5m (61ft × 5ft 3in) in 1971. The tallest was at Pencarrow, Cornwall, measuring 24m × 1.8m (79ft × 6ft) in 1970.

This has now been overtaken by several trees, as recent measurements show:-

Birr, Co. Offlay, 24.4m × 1.1m (80ft × 3ft 9in) 1987
Borde Hill, Sussex, 26.8m × 2.2m (88ft × 7ft 5in) 1989
Royal Botanic Garden, Edinburgh, 20m × 1m (66ft × 3ft 4in); 15.2m × 1.5m (50ft × 5ft) 1987
Smeddon House, E. Lothian, 15.8m × 1.2m (52ft × 4ft 3in) 1988

Saumarez Park, Guernsey, 25m × 2.2m (82ft × 7ft 6in); 20m × 1.5m (66ft × 5ft 4in) 1988

Although this pine has in general failed in coastal plantings in Western Europe, except in Ireland, it seems to be perfectly hardy on the Atlantic coast of North America, especially in New England. It withstands seashore conditions there and can even put up with salt spray.

It was introduced into the USA in 1855 and, apart from its use in coastal areas, it is popular in gardens because of its picturesque and irregular manner of growth.

Lodgepole or Shore Pine

Pinus contorta

A general description of the Lodgepole pine is difficult as its various habitats are so different and consequently its size and shape are too. The coastal variety of Shore Pine *P. contorta* var. *contorta*, for example, is short, gnarled and stunted, while the inland variety of Lodgepole Pine *P. contorta* var. *latifolia* is tall, straight and upstanding. Suffice it to say that the bark is thin and scaly and the yellow-green needles tend to hug the shoot, especially in the Shore Pine, and are twisted and much shorter than those of most pines. (See illustration facing p.145)

Distribution and habitat

The natural range of the Shore Pine stretches from south-east Alaska right down to northern California in a belt about 160 kilometres (100 miles) wide, and restricted generally to altitudes below 600m (2,000ft). The range of the Lodgepole Pine stretches from the Yukon to south east California at altitudes varying between 920 and 3,500m (3,000 and 11,500ft).

Being a tree which occupies burnt over and otherwise cleared areas, this pine is frequently found in nearly pure stands with rarely more than 20% of other species; usually Western White Pine *P. monticola*, Ponderosa Pine *P. ponderosa*, and sometimes such species as Douglas Fir *Pseudotsuga menziesii*, Jeffrey Pine *P. jeffreyi*, Englemann's Spruce *Picea englemannii* and Western Larch *Larix occidentalis*.

Life history

Cones are borne at a very early age – 5 to 10 years – and plentiful crops of seed are produced nearly every year, most cones releasing their seed in the second season but some remaining on the branches for years before liberating the seed. The huge volume of seed production often results in overstocking and stagnation with 10-year-old seedlings being little more than 15 to 25cm (6 to 10in) tall, and equally slow rooting, resulting in instability.

Sizes attained in maturity vary greatly, some Shore Pines being no more than 6 to 12m (20 to 40ft) tall and 0.5 to 1.5m (2 to 5ft) in girth. But the Lodgepole variety is usually up to 23m (75ft) at an age of about 180 years, with some individuals twice that height. The oldest recorded Lodgepole Pine was 450 years of age, growing in Montana, but there are reports of specimens living to 600 years.

These pines are intolerant of shade, although they can maintain themselves in dense, pure stands for up to 100 years, but as they require more moisture in the soil and air than Douglas Fir or Ponderosa Pine, they eventually find themselves at a disadvantage.

Timber

The light brown timber is hard and rather brittle, but being available in large quantities, it has many uses including sleepers, construction timber, poles and fuel.

Introduction, performance and use as an exotic

The Shore Pine was discovered by David Douglas near Cape Disappointment in 1825, but apparently was not introduced to Britain until 1855 when it appeared in Lawson's Nursery catalogue. The Lodgepole Pine was discovered by Jeffrey in 1852 and was introduced the following year.

Loudon describes the species and its discovery but gives no date of introduction. Elwes mentions two trees at Westonbirt which he describes as Lodgepole Pines, one of which measured, in 1909, 18m × 1.3m (59ft × 4ft 7in). Mitchell describes it as a Shore Pine which he found to measure 22m × 2.8m (72ft × 9ft 6in) in 1987. Another specimen at Grayswood, Surrey, which both authorities agree is a Shore Pine, was 8.5m × 0.9m (28ft × 3ft 1in) when measured by Elwes in 1906, and 20m × 2.6m (66ft × 8ft 9in) when Mitchell saw it in 1971.

The largest of the species so far grown in Britain was one at Bicton, Devon, which measured 31.4m × 3.3m (103ft × 11ft 3in); unfortunately it blew down in 1966. Mitchell rated it a Shore Pine, James says it was a Lodgepole.

Most of the old Shore Pines, according to Mitchell's data, are between 18 and 27m (60 and 90ft) tall, whereas the Lodgepoles, all very much younger and thus probably shorter lived in Britain, are between 9 and 21m (30 and 70ft).

Recent measurements include:-

P. c. contorta
Bodnant, Gwynedd, 33.5m × 3m (110ft × 10ft) 1988
Castle Milk, Dumfries, 24.4m × 2.2m (80ft × 7ft 7in) 1987
Wakehurst Place, Sussex, 19m × 2.2m (62ft × 7ft 6in) 1988

P. c. latifolia
Borde Hill, Sussex, 25m × 1.5m (82ft × 5ft 1in) 1987
Culcreuch Castle, Stirling, 29.5m × 2m (97ft × 7ft 2in) 1987
Pitcarmich, Perth, 15.8m × 1.5m (52ft × 5ft 4in) 1987
Westonbirt, Glos., 26.8m × 2.4m (88ft × 8ft 4in) 1987

The two varieties together, with the Shore Pine greatly in the majority, come second only to Sitka Spruce in importance as exotic forest trees in Britain because of their ability to grow quickly on poor soils and exposed sites, making them outstanding pioneer species for afforestation in upland areas. The best performers amongst the Shore Pines are those from Washington and Oregon, although their speed of growth leads to deformity in persistent strong winds. But the Lodgepole Pine, although growing straighter, is less able to compete with the heather so abundant on the sites in question. These pines are sometimes planted as pure crops but more often as nurses for less hardy trees. Natural regeneration occurs freely along ridesides.

The timber appears to have the same working properties as Scots Pine, but is not so strong. To date not enough of the home-grown product has been produced to evaluate it properly.

Lombardy poplar

Populus nigra 'Italica'

The Lombardy Poplar is a very narrow, tall, deciduous tree, often 30m (100ft) in height, with a short, rugged bole covered in ridged, dark grey bark. Masses of upright branches grow quite close to the ground and form a flame-like crown with their triangular, dark green leaves. (See illustration facing p.145)

Distribution and habitat

Although some authorities still maintain that the Lombardy Poplar is a separate species, it is generally thought to be a variety of the Black Poplar *Populus nigra*; a tree with a wide range in Europe and Western Asia. This theory has been reinforced by a report in the Kew Bulletin of 1929, which stated that a Lombardy Poplar in New Zealand reverted to the normal form.

It seems most likely that this poplar originated as a 'sport' on the banks of the River Po in Northern Italy at the beginning of the 18th century, although at one time it was thought to have come from Persia and the Himalayas. Certainly, there were, a century ago or more, some very large Lombardy Poplars in Kashmir and other parts of Asia and some are still to be found there. These and their forerunners were almost certainly planted. That its origin is comparatively recent is supported by the fact that this tree was unknown to the classical and mediaeval writers.

All the typical Lombardy Poplars, tall and narrow, are males and may well have originated from cuttings from the original tree. A female tree of a rather less fastigiate habit and not so tall exists, named *P. nigra* 'Italica Foemina'. Elwes was of the opinion that it was not a female Lombardy Poplar but a hybrid of the Black Poplar and a Lombardy. There is another cultivar, *P. nigra* 'Plantierensis', which originated in Metz in 1855; it tends to have brighter and more healthy-looking leaves than the true Lombardy, less taper and a square top.

Timber

The timber is valueless as the trunks are too short and branchy.

Introduction, performance and use as an exotic

It is generally thought that this poplar was first introduced into Britain when, in 1758, the Earl of Rochford, who was ambassador in Turin at the time, planted some cuttings on his estate at St Osyth's Priory, Essex. But Loudon mentions that Archibald, Duke of Argyll, planted one at Whitton. As he started his tree-planting there in 1720 and died in 1761, it is possible that he may have beaten Lord Rochford to it. The tree was still at Whitton in 1838 when Loudon measured it as 35m × 5.8m (115ft × 19ft 8in). However, the 1758 date is better substantiated; in 1909 Elwes received photographs of two of the St Osyth's trees from the gardener there. The larger of the two was 'a hollow stump divided into two trunks about 20ft (6m) high and measuring 18 feet (5.5m) in girth at three feet (90cm) from the ground. Living branches have sprung from different places in the trunk, two of which attain a height of about 50 feet (15m).'

Elwes, quoting the *Transactions of the Scottish Arboricultural Society* of 1879, mentions a specimen at Hampden Court, Herefords. which was said to be 49m (160ft) tall, but he found no trace of it in 1905. Loudon found none taller than 38m (125ft) (and this one was only 50 years old). The largest which Elwes knew of was at Greatbridge House, Romsey, Hants., measuring 39.6m × 4.1m (130ft × 13ft 9in). It died in 1881. Indeed, a great many succumbed to the winters of 1879-81, even more than in 1860-61. The largest in the country in 1975, according to Mitchell, was at Marble Hill, Twickenham; it was just under 30m (100ft) tall. This tree has now been superseded by one in the University Parks, Oxford, which Mitchell measured as 37m × 3.7m (121ft × 12ft 6in) in 1981. Other recent measurements include the following trees:-

Central Park, Bournemouth, 37m × 2.5m (121ft × 8ft 8in) 1988
Fawley Court, Oxon, 35m × 3.6m (115ft × 12ft 2in) 1987
Newby Hall, Yorks., 36m × 2.2m (118ft × 7ft 9in) 1987
Oxford University Parks, 33m × 4m (108ft × 13ft 5in) 1988
The Courts, Avon., 35m × 1.5m (115ft × 5ft 4in) 1987

The Lombardy Poplar is very common in Britain, except in hilly areas, and is especially so in the Thames and Severn basins. Some may be 'Plantierensis' and a few 'Italica foemina', but the majority are typical trees, planted as screens round factories, along rural roads and in larger gardens. But the tree is not long-lived and becomes very scruffy in old age.

This poplar was introduced into the USA in 1784, when W. Hamilton planted a specimen in his garden at Woodlands, near Philadelphia. At one time it was extensively planted for shelterbelts, screens and along roads, but it is so short-lived, losing its shape and succumbing to canker as early as 20 years of age, that it is not now recommended. Fastigiate forms of *P. simonii*, a Chinese species, and of the native Aspen *P. tremuloides* are taking its place.

The Lombardy Poplar is much used for formal planting and for shelterbelts and screens in New Zealand and in parts of Australia.

Black Cottonwood

Populus trichocarpa

The Black Cottonwood or Western Balsam Poplar is the largest of the American poplars and the largest broadleaved tree in North West America where it is a minor component of the great forests of the Pacific Coastal region. The stout trunk, clad in ashy-grey bark, deeply divided into broad round ridges, is often clear of branches for half its height. It supports a broad, open crown of erect branches, up to 45m (150ft) tall, bearing slightly angled twigs and thick, rather pointed oval leaves, dark green above and silvery beneath. The male and female catkins are borne on separate trees. (See illustration facing p.145)

Distribution and habitat

The natural range of the Black Cottonwood stretches from Kodiak Island and southeast Alaska in the north, including nearly all British Columbia and south west Alberta, and south along the Pacific Coast to the mountains of southern California; east into Montana, central Idaho and north Utah. The trees reach their maximum development in the coastal areas on deep alluvial soils in forest meadows and along river banks. They also grow well, but to smaller sizes, in mountain areas as high up as 2,140m (7,000ft) in protected valleys and canyon bottoms.

In Alaska and Yukon it merges with the Balsam Popular *P. tacamahaca* and in Alberta with the Plains Cottonwood *P. deltoides*; but for the greater part of its range it consorts with the great conifers – Douglas Fir *Pseudotsuga menziesii*, Western Red Cedar *Thuya plicata*, Western Larch

Larix occidentalis and Western Hemlock *Tsuga heterophylla*; and at higher latitudes with Sub-alpine Fir *Abies lasiocarpa* and Englemann's Spruce *Picea englemannii*.

Life history

The catkins of the Black Cottonwood appear before the leaves, and the female trees start to drop their seeds in May and June. The seeds are dispersed over a wide area by the wind and germinate almost immediately if they fall in favourable damp sites. Growth is rapid after the crowded mass of seedlings have thinned themselves naturally during the first five years. On ideal sites, such as occur in the Puget Sound area, height growth of 15m (50ft) in 10 years is not unusual. Reproduction by cuttings is quick and easy.

Although giant specimens over 60m (200ft) tall and 5.4m (18ft) in girth were once to be seen in the coastal areas, they are now nowhere to be found. Nevertheless, these poplars grow so quickly that sizeable trees are numerous and increasing in number. Among some selected specimens in the Fraser Valley in British Columbia, measured recently, was one which had grown to 36.6m × 2.5m (120ft × 8ft 6in) in only 27 years; and another which, in 17 years, had reached 29m × 1.5m (95ft × 5ft).

Such speedy growth suggests a relatively short life span. Maturity is attained in about 75 years, although trees of up to 200 years have been reported, but at this advanced age they are usually badly damaged by storm and snow.

This poplar is very intolerant of shade but its rapid growth often gives it a period of dominance. Large trees are usually on the forest edges.

Timber

The timber is greyish-white, soft, weak and far from durable. But it can be obtained in large sizes making it useful for plywood and veneers, as well as for boxes and fuel.

Introduction, performance and use as an exotic

Surprisingly, the Black Cottonwood was not grown in Britain until a tree was planted in the Edinburgh Botanic Garden in 1892. This tree was 11.8m × 0.7m (39ft × 2ft 6in) when Elwes measured it in 1906. Another specimen at Kew, planted in 1896, was 16.7m × 1.1m (55ft × 3ft 10in) in 1911. A tree planted by Elwes himself at Colesbourne in 1903 was 9.7m (32ft) tall eight years later. But the early optimism which this rapid growth encouraged was later dispelled by the tree's susceptibility to canker, for these early plantings were nearly all from the same clone and thus equally prone to infection. Nevertheless, there are some large specimens from later plantings to be found in Britain. Recent measurements include some of them:-

Bowood, Wilts., 41m × 3m (135ft × 10ft 4in) 1988

Colesbourne, Glos., 36m × 2.5m (118ft × 8ft 6in) 1987
Thorpe Perrow, Yorks., 33.5m × 1.9m (110ft × 6ft 6in) 1987
University Botanic Gardens, Camb., 34m × 3.3m (112ft × 11ft) 1988
Wester Elchres, Inverness, 41m × 2.5m (135ft × 8ft 7in); 40.2m × 2.6m (132ft × 8ft 10in) 1988

A hybrid between the Black Cottonwood and the Balsam Poplar *P., tacamahaca*, known as TT32, has proved to be very vigorous and less susceptible to canker. Another hybrid, this time with an Asiatic species, *P. maximowiczii*, called P. 'Androscoggin', is probably the quickest growing tree now cultivated in Britain – 15m (50ft) in five years – and, as more and more resistant strains are bred, should have a great future.

These and other Balsam Poplar hybrids are more satisfactory growers in western Britain than the Black (European) Poplar hybrids.

Weeping Willow

Salix babylonica

The Weeping Willow is a small tree, 9 to 12m (30 to 40ft) tall, with a short trunk clad in rough, ridged bark and bearing wide-spreading, ascending branches. The long ends of the branches recurve and descend perpendicularly, the thin, trailing brown twigs reaching the ground in cascades of deep green, lance-shaped leaves. (See illustration facing p.160)

Distribution and habitat

The Weeping Willow is a native of China where it is found growing naturally in the mountainous region of the Upper Yantgse River as well as in the foothills surrounding the Great Plains of the northern coastal regions, usually near water courses and in company with other species such as *Salix matsudana* and *Populus tomentosa*. Since ancient times this willow has been associated with the Chinese people, playing an important part in their gardens, folklore and painting, and has thus been planted throughout China. In Japan too, where it was introduced in early times, it is extensively planted.

Life history

The male and female catkins, usually on separate trees but sometimes, rare for a willow, on the same tree, expand in March, the seeds, the 'willow down' so common in China, being dispersed by the wind. Although in the wild natural seedlings do appear, reproduction is normally effected by cuttings which root readily and grow very rapidly.

Timber

The timber is of no commercial value.

Introduction, performance and use as an exotic

The Weeping Willow probably came out of China towards the west at a very early date, as willow branches were used extensively for making baskets and crates carried by merchants travelling along the 'silk' route. The twigs root readily when purposely or accidentally stuck in the ground; in this way the willow became established in Central Asia and Asia Minor, and perhaps even in Greece at the end of the 17th century. It appears to have been introduced into Britain some time before 1730 by a Mr Vernon, a Turkey merchant from Aleppo, who planted it in his Twickenham Park garden near London. This specimen is said to have come from the banks of the Euphrates, and was no doubt a descendant of those trees brought from China by the old silk merchants. The presence of Weeping Willows in that part of Asia, albeit introduced, and the verse from Psalm 137 'By the waters of Babylon, there we sat down, yea we wept, when we remembered Zion. We hanged our harps upon the willows in the midst thereof' no doubt account for the scientific name given to this willow by Linnaeus. But the trees upon which the harps were hung were probably Poplars *Populus euphratica*.

A tree so easily propagated by cuttings is bound to have had many chance beginnings such as the one grown by Pope from a withy binding a package received by Lady Suffolk from Spain. The Empress of Russia took cuttings from it in 1789. It was felled in 1801 and many trinkets and ornaments were made from its timber. Another famous Weeping Willow was one planted by Governor Beatson on St Helena in 1810, which became a great favourite of Napoleon who often sat beneath it. Many cuttings from it were sent to Britain after the Emperor's death in 1821, and the tree continued to thrive at least until 1912 when it measured 11m × 1.3m (36ft × 4ft 6in).

Loudon records that there were many Weeping Willows between 15 and 18m (50 and 60ft) tall growing on the banks of the Thames in 1838. Elwes thought it unlikely that any of these were still alive at the turn of the century. He recorded and photographed a specimen, planted at Cheltenham in 1860, which measured 22.8m × 2.7m (75ft × 9ft) in 1911.

It is these considerable sizes, so much greater than those of the natural trees in China, and the now established fact that many Weeping Willows are hybrids of *Salix babylonica* with other species, which cast doubt on the true identity of the trees mentioned by both Loudon and Elwes.

The weeping variety of the White Willow *S. alba* 'Tristis', thought by some to be a cross between *S. babylonica* and the golden form of *S. alba*, occurred in France in about 1800 and is now one of the commonest weeping willows in Britain, being much hardier and taller growing than *S. babylonica*.

Recent measurements of *S. alba* 'Tristis' include the following:-

Huntington (town), 18.3m × 3.5m (60ft × 11ft 10in) 1988
Oxford Botanic Gardens, 23m × 3.7m (76ft × 12ft 5in) 1987
Radnor Gardens, Twickenham, 25m × 4.2m (82ft × 14ft) 1987
Regents Park, London, 22.8m × 3.6m (75ft × 12ft 3in) (at 30cm (1ft)) (1981)
West Kingsclere Farm, Hants., 18.3m × 5.2m (60ft × 17ft 8in) 1987

In the USA, *S. babylonica* is a popular ornamental tree in gardens and parks in the warmer regions. It is replaced in the colder parts by hybrids between *S. babylonica* and the European Crack Willow *S. fragilis*, namely the Thurlow Weeping Willow *S.* x *elegantissima*, and the Wisconsin Weeping Willow *S.* x *blanda*; and by another hybrid between *S. alba* and *S. babylonica* named *S.* x *sepulcralis*. This latter tree is also grown in Britain and is perhaps the finest and most vigorous weeping willow found here.

All the weeping willows, the true species and the hybrids, make fine subjects for garden and park planting, especially beside lakes and ponds, although their rapid growth can cause embarrassment in small gardens. They leaf earlier than any other deciduous trees, their pale greens being particularly welcome in the grey days of early March; and they have an elegance scarcely challenged by other trees.

Caucasian Wingnut

Pterocarya fraxinifolia

The Caucasian Wingnut is a large tree from the Caucasus Mountains and northern Persia. It has a short bole with a dark grey, furrowed bark. Large branches emerge low down on the trunk and curve upwards to form a wineglass-shaped or sometimes a very spreading crown of long, 25cm (10in), pinnate leaves with 20 to 25 leaflets. (See illustration facing p.160)

Distribution and habitat

The natural range of this wingnut stretches from the lower slopes of the Caucasus Mountains up to 4,870m (1,600ft) along the eastern shores of the Black Sea, the southern shores of the Caspian Sea and into the northern provinces of Iran. It is usually found growing in damp places in company with Alders *Alnus glutinosa* in the lower lying areas, and on well-watered soils in deciduous forest with Oriental Beech *Fagus orientalis*, Eastern Hornbeam *Carpinus orientalis* and Sessile Oak *Quercus petraea*.

Life history

The male and female flowers, in catkins, open in April and May, the females on slender hanging stalks which lengthen to as much as 45cm (18in) as the flowers, strung along it develop into little nuts, completely surrounded with whitish-green wings.

The seedlings grow very quickly, producing soft, pithy shoots up to 60cm (2ft) long in two seasons. They can be as much as 9m (30ft) tall in 10 years, with an ultimate height in maturity of 30m (100ft) and a girth of 3m (10ft).

This tree also reproduces itself by root suckers and is often surrounded by a mass of such suckers of varying heights.

Timber

The soft, light timber, very like that of lime trees in appearance and texture, is used locally for making boxes and other general uses, but is little exploited commercially.

Introduction, performance and use as an exotic

The Caucasian Wingnut first came to Europe when Michaux introduced it into France in 1782. It was not brought to Britain until some time in the early 1800's.

Loudon mentions a tree he saw at Croome, Worcs., which in 1838 was 15 years old and 7.5m (25ft) tall. But Henry, writing about 70 years later, reported that it was long since dead.

The specimen which Elwes photographed and measured at Melbury, Dorset, was 27.4m × 3.3m (90ft × 11ft) in 1906. It still lives and in 1989 it measured 36m × 5.7m (118ft × 19ft 2in). A specimen at Claremont Park, Surrey, which Elwes also photographed, was 15.2m × 5.4m (50ft × 18ft) in 1903, but the bole was divided into eight limbs; even then the heart was beginning to decay and would not have long been able to support the huge crown which was 27m (90ft) in diameter.

The Melbury tree remains the finest in the country, and there are many others in large gardens and collections up and down the land. But this tree is not easy to establish because the soft shoots often do not ripen sufficiently before the first frosts and are damaged then or later by the early spring frosts. Damp soil or a site near open water seems to be essential for good growth.

Recent measurements include the following trees:-

Abbotsbury, Dorset, 37.8m × 3.9m (124ft × 13ft 2in); 32m × 5.2m (105ft × 17ft 1in) 1988
Hyde Park, London, 21.3m × 4m (70ft × 13ft 4in) 1987
Lacock Abbey, Wilts., 26m × 5.9m (85ft × 19ft 10in) 1987
Powerscourt, Co. Wicklow, 26.8m × 3.3m (88ft × 11ft 4in) 1988

The Caucasian Wingnut was introduced into the USA in 1850, but is not commonly planted as an ornamental, although in some collections it claims interest because of its long racemes of fruit in the summer time. It is interesting to note that Wyman describes its habit as 'wide-spreading, open', a shape it rarely adopts in Britain.

Common Walnut

Juglans regia

The Common Walnut is a large, deciduous tree with a huge, spreading crown of thick, sinuous branches, supported by a massive trunk clad in pale grey, deeply fissured bark, smooth between the fissures. The dark, yellow-green compound leaves are very large (up to 45cm (18in) long). (See illustration facing p.160)

Distribution and habitat

This walnut has a very wide distribution in the northern hemisphere extending from Greece, its western limit, through the Balkans to Asia Minor, the Caucasus, Iran and into the Himalayas, Burma and Central China.

In Europe it is usually in mixture with oaks and chestnuts and, in moist valleys, with Silver Fir *Abies alba* at altitudes between 610 and 1,310m (2,000 and 4,300ft). But Henry says that small, pure stands, undoubtedly wild, occurred on the northern slopes (which are rich in springs) of the mountains of what is now Yugoslavia, at altitudes around 610m (2,000ft). In the Caucasus the tree grows up to 1,370m (4,500ft) and in Afghanistan and the Himalayas as high up as 3,050m (10,000ft).

In the Himalayas it is a minor component of the temperate and alpine broadleaved forest regions, occurring in mixture with such species as Indian Horse Chestnut *Aesculus indica* and Turkish Hazel *Corylus colurna*, as well as with various evergreen oaks such as *Quercus dilatata*. It also appears in the alpine coniferous forests in glades and gulleys, together with *Sorbus cuspidata, Betula utilis* and *Acer* spp. as well as many others in an understorey beneath Morinda and Sikkim Spruces *Picea smithiana* and *P. spinulosa* and Bhutan Pine *Pinus wallichiana*.

In China it is found wild in the essentially evergreen broadleaved forest of the warm and humid valleys of South Sikiang and North Yunnan in mixture with evergreen oaks *Castanopsis* spp. and maples such as *Acer davidii* and the evergreen *A. oblongum*, as well as with the conifers *Torreya fargesii* and *Taiwania cryptomerioides*.

Some authorities once maintained that the Common Walnut was native to Japan where it is cultivated as an ornamental and as a fruit tree, but it seems more likely that it was introduced from China in early times.

Life history

The female flowers, in bunches of five or sometimes singly, appear on the ends of the young shoots in late May and develop into round green fruits each containing a wrinkled-shelled nut. The seedlings grow vigorously and attain some 9m (30ft) in height in about 10 years when they start to produce fruit, all the while developing an extremely tenacious and deep-penetrating root system rendering even the largest trees almost immune to windthrow. Although in some fertile valleys wild trees may reach top heights of 30m (100ft), the walnut rarely exceeds 21m (70ft) and does not appear to be long-lived, 200 years being about the maximum age.

Timber

Before the advent of mahogany in the 1750's, walnut timber was the most valuable wood for furniture making. It is very variable in colour – greyish brown with smoky streaks – hard, tough and seasons well, if slowly. It is now mainly used for veneers, but also for gunstocks and fancy goods such as fruit bowls etc. Homegrown logs of any size are scarce, the main supplies of planks coming from France and Italy. In 1906 Elwes saw a walnut log in France which was 8.5m (28ft) long and 4.5m (15ft) in girth, containing 15.5cu. metres (560cu.ft) of timber. A tree cut near Lausanne in 1900 took 24 horses to haul it. It contained 19.5cu. metres (700cu.ft) of timber and was sold for gunstocks for £150. Such logs as these must now be rare indeed.

Introduction, performance and use as an exotic

It is generally accepted, although there is no real evidence to support the theory, that the Walnut was introduced into Britain by the Romans, who cultivated it here for its nuts.

One of the trees, mentioned by Loudon, at Finborough Hall, Suffolk, was 30m (110ft) tall and only 100 years old, but none of the others he listed exceeded 21m (70ft) in height. The specimen he measured at Melbury, Dorset was said to be 200 years old but was only 20m (66ft) tall. Elwes gives details of a 'record' walnut (taken from early *Transactions of the Royal English Horticultural Society*) which once grew at Lamport Hall, Northants., and which, in 1884, was sold to Birmingham gunmakers. It contained 23.1cu.m (816cu.ft) of sound wood of which the butt, 3.6m (12ft) long and 5.4m (18ft) in girth, accounted for 6.9cu.m (243cu.ft) and one limb for 3.1cu.m (108cu.ft). In 1905 Henry measured a tree, at Laverstock Park, Hants., as being 24.4m × 4m (80ft × 13ft 8in); and this same tree had grown to 25m × 5m (82ft × 17ft 2in) when measured by Mitchell in 1957. A tree at Gayhurst, Newport Pagnell, Bucks., admired by Henry and measured in 1904 as 24.4m × 5m (80ft × 17ft), was photographed by Gardner in 1949 when it girthed 5.7m (19ft) (*Quarterly Journal of Forestry* 55:4). It was measured again by Maynard Greville in 1955 when it was 21.3m × 5.8m (70ft × 19ft 9in). In 1964 it came under threat from veneer hunters and, to save it, the Buckinghamshire County Council bought it and the land it stood on for £1,500. A photograph appears in the July 1965 issue of *Quarterly Journal of Forestry*. In 1976, Mitchell found this tree to be 23.4m × 6m (77ft × 20ft).

The *Journal* (19:4) reported a tree at Cuxham, Oxford, as measuring 32.3m × 5.1m (106ft × 17ft 4in) with a spread of 34m (112ft), and the issue 58:3 publishes a photograph of a well propped-up walnut at Chepstow Castle, reputed by the keeper to be 600 years old, but no proof or evidence for this claim is given.

The Walnut with the largest girth found by Mitchell was at Piltdown, Northants. which measured 6.5m (21ft 6in) in 1955 Other measurements include the following trees:-

Cirencester Abbey, Glos., 26m × 3.2m (85ft × 10ft 7in) 1988
Fingask, Perth, 28m × 3.6m (92ft × 12ft 2in) 1987
Garth House, Perth, 29.8m × 3.6m (98ft × 12ft 3in) 1988
Lydham Hall, Salop, 22.8m × 3.9m (75ft × 13ft 3in) 1987
Preston Hall, Middlesex, 24.4m × 3.9m (80ft × 13ft 3in) 1987
Strafford Park, Glos., 21.3m × 3.3m (70ft × 11ft) 1987

The Common Walnut is rarely planted solely for the production of timber and only one such plantation has so far been successful. It was established on old forest land in south-west England at an elevation of 120m (400ft) during 1937-40. After 20 years it had reached a top height of 8.5m (28ft). This tree is more often planted for its nuts or as an ornamental, or both, although for nut production budded clones are more satisfactory. For ornamental planting alone the cultivar 'Laciniata', with cut leaves, is an attractive alternative.

The Common, or Persian Walnut as it is called in the USA, was introduced there in early colonial times. The type tree there is mainly cultivated for its nuts: over 50,600 hectares (125,000 acres) of such orchards in California alone. For ornamental planting the cut-leaved cultivar is more favoured and some clones of this are hardy as far north as Boston. A cultivar 'Carpathian' introduced into Canada from Poland in 1925, has withstood temperatures as low as -40C (-40F) and thus shows promise in some of the colder areas.

Black Walnut

Juglans nigra

The Black Walnut is one of the largest deciduous trees growing in the eastern half of the USA, and parts of eastern Canada. The rather open, round-headed crown of 60cm (2ft) long, shiny, compound leaves is supported aloft 30m (100ft) and more by a sturdy bole clad in dark brown, deeply fissured bark, and by large, upward-growing somewhat twisted boughs. In the dense forest the bole can be free of branches for up to 18m (60ft), with little or no taper. (See illustration facing p.160)

Distribution and habitat

The natural range of the Black Walnut stretches west from Maine to Minnesota, and south to Louisiana and Florida, with the best trees in sheltered valleys of the Appalachians and the Mid-West.

This tree is very sensitive to soil conditions which, for good growth, must be deep, well drained and nearly neutral. In the Appalachians it is rarely found above 1,220m (4,000ft), the best trees in the mountains growing along streams and at the base of north and east facing slopes.

Some small pure stands occur, usually on the edges of forests, but generally this walnut is scattered through the woodland together with, in the north, Black Cherry *Prunus serotina*, Eastern Red Cedar *Thuya occidentalis*, Green Ash *Fraxinus pennsylvanica* and Basswood *Tilia americana*; whilst further south White Elm *Ulmus americana*, Box Elder *Acer negundo* and Hackberry *Celtis occidentalis* are more common associates.

Life history

Although some seed production can start at the early age of eight years, the best production occurs after 30 years. The fruits, borne in clusters, are globular, light green and 5cm (2in) in diameter, containing the hard nut with the typical grooved shell. They are disseminated by gravity and animals, and usually germinate in the following spring, but sometimes a year later. Seedlings grow up to 90cm (3ft) in the first year and to as much as 1.8m (6ft) in the second; thereafter 60 to 90cm (2 to 3ft) annually reaching 9 to 12m (30 to 40ft) in 20 years, with girths of 45 to 60cm (1ft 6in to 2ft). They mature at 150 years but some may go on to 250 years when they can be 45m (150ft) tall and 5.4m (18ft) in girth.

The Black Walnut is intolerant of shade and never becomes a climax species, usually occurring as individuals beneath openings in the canopy.

Timber

The soft brown, easily worked timber is strong and shock-resistant. When large trees were readily available this fine figured wood was (and still is) in great demand for furniture and cabinet making. Traditionally used for aircraft propellers, it is also used for gunstocks and other things where strength and shock resistance are required.

Introduction, performance and use as an exotic

The Black Walnut was probably one of the first American trees to be cultivated in Britain. Tradescant, Junior, was growing it before 1656 and Bishop Compton had a specimen in his Fulham garden in 1788. This

Weeping Willow, Regents Park, London (p.153)

Caucasian Wingnut, Abbotsbury, Dorset (p.155)

Common Walnut, Dancer's End, Tring, Herts. (p.157)

Black Walnut, Bishop's Palace, Wells, Somerset (p.159)

Himalayan Birch, R.B.G., Kew (p.162)

Cherry Birch, Lizard Hill, Weston Park, Staffs. (p.163)

Roble, Stowe School, Bucks. (p.165)

Raoul, Howick Hall, Northumberland (p.167)

tree, according to Loudon, was 15.2m × 1.5m (50ft × 5ft) in 1838. The fine tree at Syon House, mentioned by Loudon as being 24m × 0.9m (79ft × 2ft 11in), had lost its top when Elwes saw it in 1903, but girthed 3m (10ft).

Elwes gives a photograph of a Black Walnut at Twickenham which, in 1905, was 29.8m × 4.2m (98ft × 14ft 3in), with a spread of 28m (93ft). Some of these trees planted at Stratsfield Saye, Hants., in 1885 were 3m (10ft) tall when Elwes saw them 18 years later, and one of them girthed 1.8m (5ft 11in) when Gardner measured it in 1946. Another tree at Corsham Court, Wilts., which girthed 3.3m (11ft) in 1903, had increased to 3.7m (12ft 5in) when Gardner saw it in 1947. By 1989, when Mitchell last measured it, it stood at 26m × 4.3m (85ft × 14ft 5in).

Elwes also photographed a specimen at Mote, near Maidstone, which was planted around 1800 and measured 31m × 3.6m (103ft × 12ft) in 1902. By 1977 it had grown no taller but had attained the huge girth of 5.7m (19ft 3in).

There are still many very large Black Walnuts in England, mostly in the Thames Valley and East Anglia. The tallest specimen recorded by Mitchell, from Maynard Greville's 1952 measurements, is one at Hatfield Forest, Essex, standing at 32m × 4.2m (105ft × 14ft). A tree at Battersea Park, London had reached this height when measured by Mitchell in 1979, but 10 years later it had grown to 33.8m × 3m (111ft × 10ft 5in). Another at Syon House, Middlesex girthed 4.9m (16ft 8in) in 1961.

Recent measurements include the following trees:-

Albury House, Surrey, 26m × 4.7m (85ft × 15ft 10in) 1988
Cambridge Botanic Gardens, 21.3m × 3.3m (70ft × 11ft 2in) 1987
Corsham Court, Wilts., 26m × 4.3m (85ft × 14ft 5in) 1987
Much Hadham, Herts., 36m × 6.4m (118ft × 21ft) 1987
Oxford Botanic Garden, 31m × 3.3m (102ft × 11ft 4in) 1987
The Mote, Kent, 29m × 5.8m (95ft × 19ft 7in) 1987

As a forest tree, the Black Walnut has rarely been tried in Britain. In a mixed plantation with oak at Spixworth Hall, Norfolk, planted in 1830 or thereabouts, Black Walnut, after 80 years, accounted for the same volume as the oak 43 to 49cu.m/ha (635 to 720cu.ft/ac)and girthed about 1.5m (5ft). Some of the flooring at Spixworth Hall was made from them. But, generally speaking, the Black Walnut has no place in British forestry, and the tree is planted only as an ornamental. It is perhaps more handsome than the European species *J. regia*, but is not as hardy and the nuts are quite inedible.

This native walnut is not a popular ornamental in the USA because it is too coarse for street planting and the nuts can cause a road hazard. It has also been proved (according to Wyman) that the roots give off a substance toxic to other plants and thus it should not be grown near valued specimens. The cut-leaved variety 'Laciniata' is more often planted.

Himalayan Birch

Betula utilis

In that most notable characteristic of birches, the colour of their bark, the Himalayan Birch is very variable. Some trees have mainly white bark, peeling off in strips; in others it is orange-red, especially in young trees, becoming darker and thicker at the base of the bole with age. This bole supports a rounded crown of ascending branches bearing shiny, dark green, oval leaves, somewhat larger than the European species. (See illustration facing p.161)

Distribution and habitat

The several forms (or species) of Asiatic birches merge with one another so gradually that exact distribution is difficult to define. The natural range of *Betula utilis* seems to stretch all along the Himalayas from Kashmir to Sikkim, and on into western and central China where it was once considered to be *Betula utilis* var. *albo-sinensis*. It is now named as a separate species *Betula albo-sinensis*, its chief characteristic being the peeling orange-red bark, the whiteness continuing up into the larger branches. This tree used to be called *Betula utilis* var. *jacquemontii*, but this also is now considered a separate species *Betula jacquemontii*.

All these trees are essentially sub-alpine, occurring between 3,050 to 4,250m (10,000 and 14,000ft), and usually the largest deciduous trees at these altitudes where, in India, they are associated with Himalayan Fir *Abies spectabilis*, Sikkim Larch *Larix griffithiana*, *Rhododendron campanulatum* and Bird Cherry *Prunus padus*. In China common associates are Sargent Spruce *Picea brachytyla* and Silver Fir *Abies delavayi* var. *fabri*.

Life history

The catkins open in May and the 3.5cm (1.5in) long females ripen in September producing the masses of tiny winged nutlets necessary to maintain the species in a hostile environment.

Growth is slow – 15 rings to 2.5cm (1in) of radius – and, although the tree eventually reaches considerable girths of up to 3.6m (12ft), heights of more than 15m (50ft) are rare, except in some dense, drawn up stands. Mostly these birches are of a gnarled and stunted habit due to regular winter snow-break.

Timber

The timber, which is hard and close-grained, is extensively used for firewood; and the bark for making paper.

Introduction, performance and use as an exotic

The Himalayan Birch was introduced into Britain by Sir Joseph Hooker in 1849. This was probably the reddish barked tree from the Himalayas. The white barked *B. jacquemontii* was introduced in 1880, and the reddish-orange barked tree from China, *B. albo-sinensis*, was brought home by Veitch in 1901.

Elwes records only one specimen, a tree at Grayswood, Surrey, which was planted in 1882 and measured 9m × 0.6m (30ft × 2ft 1in) in 1906. This was almost certainly one of the earlier *B. utilis* introductions, with reddish bark, for he does not mention this species as having bark which varies in any way from this colour.

That these birches are infrequently seen in Britain is probably due to their being very easily damaged or even killed by spring frosts, for the beauty of their bark is beyond doubt. When grown in good conditions, free from the competition and harsh environment of their natural habitat, they make fine trees between 15 and 18m (50 and 60ft) tall. Planted on lawns or in borders amongst shrubs, they provide a year-round display.

More recent measurements include the following trees:-

Crarae, Argyll, 15.2m × 1.2m (50ft × 4ft) 1976
Howick Hall, Northumberland, 18.3m × 1.2m (60ft × 4ft 2in) 1978
Hyde Park, London, 19m × 1.1m (62ft × 3ft 7in) 1988
Trewithin, Cornwall, 17m × 1.5m (56ft × 5ft 3in) 1985

The Himalayan Birch as such does not appear to be much planted in the USA, but the closely related (and some say only sub-species) Chinese Paperbark Birch *B. albo-sinensis* was introduced in 1901 and is planted for its winter display of orange-coloured bark.

Sweet or Cherry Birch

Betula lenta

The Cherry Birch is a tree of the north-eastern USA and Canada. It is round-topped, of graceful symmetry and, unlike most birches, has dark purplish, almost black bark which, except in the oldest trees when it becomes cracked and deeply furrowed, is smooth with horizontal corky strips like that of cherries. The deeply veined, oval 6 to 15cm (2.5 to 6in) long leaves in the domed crown are borne, somewhat sparsely, in pairs on the slender twigs. (See illustration facing p.161)

Distribution and habitat

The natural range of this birch extends from Nova Scotia, Quebec and Ontario in Canada, down the eastern United States from Maine to

Pennsylvania and along the Appalachians to north Alabama and Georgia.

It grows best on moist, well-drained soils but occurs on a wide variety of less favourable types – dry, shallow and rocky – although on some of them, despite early colonisation, it is quickly replaced by oaks and conifers. In the north of its range, in New England, it is found near sea level on the coast, but in the south it may grow at altitudes of up to 1,370m (4,500ft).

Common associates in the north are White Pine *Pinus strobus*, Northern Red Oak *Quercus rubra*, Eastern Hemlock *Tsuga canadensis* and Yellow Birch *Betula lutea*; further south such species as Basswood *Tilia americana*, White Ash *Fraxinus americana* and Tulip Tree *Liriodendron tulipifera* are found growing with it.

Life history

Seed production, in cone-like catkins 2.5cm (1in) long, begins when the tree is about 40 years old, and good crops occur nearly every year. The seeds germinate for the most part in the spring and seedling growth is fairly rapid, especially where there is light overhead shade to prevent drying out. But where, as is most often the case, they are beneath a mixed stand, growth is not so quick – something like 1.8m (6ft) in 12 years, and reaching girths of 30cm (12in) or so in 20 years. On best sites, particularly those in the valleys of South Carolina, Cherry Birch can grow up to 24m (80ft) tall in 100 years, with girths of up to 1.2m (4ft). In dense stands the trunks can be long and clear of branches, but when opened up, forking and the appearance of epicormics soon spoil their form.

For birch this species is fairly long-lived; trees of 192 and 265 years are on record.

Timber

The reddish-brown timber is close-grained, strong, hard and heavy. It is used for turnery, boxes, crates and woodenware. Birch beer is brewed from the sap, and wood alcohol is distilled from the bark and twigs.

Introduction, performance and use as an exotic

The Cherry Birch was introduced into Britain in 1759, and in 1818 the committee of the Caledonian Horticultural Society recommended its planting in deep soils of the Highland Glens. Nothing came of it.

Loudon reported a tree in Ireland which had grown to 15.8m × 0.5m (52ft × 1ft 9in), with a crown spread of 13m (42ft) in 50 years; and another 'considerable tree' at Syon House, Middlesex of which he gave no measurements. At that time, 1838, small plants were selling at 1s. 6d (8p) each. But not many seem to have been planted, or if they were, few survived. Elwes knew of only a small number: one at Oakly Park, Shropshire, which measured 18.3m × 1.3m (60ft × 4ft 9in) in 1908, which must have been exceptional; at Bicton, Devon, 11.5m × 1m (38ft × 3ft 5in) (not in James' list); and at High Canons, Herts., measuring 11m × 1.2m (36ft × 4ft 2in) in 1907.

Certainly it is a rare tree nowadays, found only in large collections. It seems to be too slow-growing and forks too readily to compete in any way as a forest tree with native birches in Britain. A plot at Bedgebury, Kent, planted in 1935, was 9m (30ft) tall 20 years later, which was not bad going; but trials at Kielder in Northumberland proved a failure.

Recent measurements include the following trees:-

Brocklesby Park, Lincs., 17m × 1.1m (56ft × 3ft 8in) 1977
Royal Botanic Garden, Edinburgh, 12.8m × 0.9m (42ft × 3ft 2in) 1981
Mt Usher, Co. Wicklow, 17.6m × 1.6m (58ft × 5ft 7in) 1989

This native birch is often planted in the USA as an ornamental, chiefly for its attractive bark and also for its autumn colouring which excels that of all other birches.

Roble

Nothofagus obliqua

The Roble, so called because the early Spanish settlers in South America thought it similar to their oaks, is a large deciduous tree with a crown of arching branches bearing small, dark green, deeply veined leaves on shoots of a regular fishbone-like pattern. The bole is clad in smooth grey bark which, with age, becomes fissured and cracked into wide, curving plates. (See illustration facing p.161)

Distribution and habitat

This tree is a southern hemisphere species found only in Chile, on the coastal ranges up to 1,520m (5,000ft), and in the Central Valley. Its most northerly limit, where it is stunted and not typical (Darwin failed to notice it when he visited the area in 1839) is just south of the capital, Santiago, at about 34°S. Roble forest proper begins to appear on the banks of the River Maule, at 36°S, from where it descends into the Central Valley and finally disappears around 41°S. Much of the Central Valley forests have been destroyed, for the Roble yields excellent timber, and they have given way to agriculture. At higher elevations the Roble consorts with *Nothofagus pumilio* and *N. antarctica*. In the Central Valley *N. dombeyi* and *Podocarpus salignus* are common associates, with N. glauca, known locally as 'Roble del Maule', as a minor component of the forest in the area of that river. The undergrowth in the main forest region is usually dense thickets of bamboo, with such shrubs as *Drymis winteri* and the beautiful *Embothrium coccineum* in the wetter more open places.

Life history

The fruits, one at the base of each leaf, are smaller but very similar to those of the European Beech *Fagus sylvatica*, and usually contain three nutlets of a size which gives about 50,000 to the 450g (1lb). In favourable conditions seedling growth is rapid, with annual height increases in the sapling stage in the order of 1.5m (5ft) a year. In maturity the Roble is never a giant tree, rarely exceeding 30m (100ft) in height, although Desmond Clarke measured one near Lake Pangipulli, in an exceptionally moist part of the forest, as 36m (120ft) tall. Old trees often girth 6 to 9m (20 to 30ft).

Timber

The dark reddish-brown timber is hard and close-grained, with very little evidence of annual rings. Native timber is excellent for construction work and is very resistant to decay. It is also suitable for turnery and for pulping. Timber grown in Britain has yet to prove itself but it seems likely that it will not match native timber in all its attributes.

Introduction, performance and use as an exotic

The Roble was introduced into England by Lobb in 1849, but this introduction seems to have been a failure for, although in 1850 plants were said to be growing well in Veitch's Exeter nursery, none of these appears to have survived even as long as the turn of the century. Elwes brought back more seed in 1902 and a tree from this seed, planted at Kew was 2.5m (8ft) tall in 1906 and 20.5m (60ft) in 1927. Mitchell records two other specimens from the Elwes seed: one at Tortworth, Glos., which seeds profusely and measured 26m × 2.4m (85ft × 8ft) in 1980; and another at Grayswood Hill, Surrey, which was 24.4m × 2.7m (80ft × 9ft 3in) in 1982.

There was a further important introduction of seed by H.F. Comber's Andean Expedition of 1925-6, which was distributed to a list of subscribers, and from this seed the notable specimens at Muncaster Castle, Cumbria, are good examples.

More recent measurements include the following trees:-

Benmore, Argyll, 29m × 1.8m (95ft × 6ft 3in) 1987
Caerhays Castle, Cornwall, 28m × 3m (92ft × 10ft) 1987
Killerton, Devon, 24.4m × 2.3m (80ft × 7ft 9in) 1983
Muncaster Castle, Cumbria, 35m × 1.5m (115ft × 5ft 3in) 1987
Stowe, Bucks., 26m × 2.3m (85ft × 7ft 9in) 1988
Trewithin, Cornwall, 26m × 2.7m (85ft × 9ft 1in) 1987

As an ornamental tree, the Roble has several merits: it grows well on poor soils, the arching crown gives it an unusual elegance, the gold and red colouring of its leaves in autumn are a delight and, of course, it grows quickly.

As a forest tree, the Roble has yet to prove itself, but it does perform spectacularly well on poor soils (as long as they are not lime-rich) where most of our native broadleaves are unproductive. It is not suitable for exposed sites, even at low elevations, as strong, cold winds cause considerable shoot dieback. It is also susceptible to early and late frosts. Nevertheless, apart from some poplars and perhaps Sweet Chestnut *Castanea sativa*, both of which are more exacting in their soil requirements, the Roble is the fastest growing broadleaved species, with a timber producing potential, that we have in Britain; and it seems to be one of the very few smooth barked hardwoods immune to bark-stripping by grey squirrels. Greater attention to provenances may well help towards finding a more frost-hardy strain.

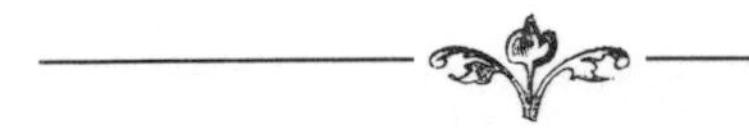

Raoul

Nothofagus procera

The Raoul is a similar-sized deciduous tree to the Roble but with a more conical profile of steeply upswept branches, bearing neat, heavily veined hornbeam-like leaves. The bole is clad in smooth grey bark which cracks vertically with age. (See illustration facing p.161)

Distribution and habitat

The natural range of the Raoul is similar to that of the Roble, but it does not extend quite so far south. The tree first appears on the River Maule at about 36°S, where, together with *Nothofagus dombeyi*, it is a minor component in the Roble forest. Between 38°S and 40°S the Raoul is at its optimum development, especially between the Renaico and Malleco Rivers, although it never forms pure stands and is always in mixture with *N. dombeyi*. South of this area it disappears after straying just over the Argentine border. The relationship between the deciduous Raoul and the evergreen *N. dombeyi* works against the ultimate survival of the former which, although able to compete in the early stages of growth, is eventually over-topped and suppressed by its evergreen companion. But in areas of regeneration, following windblow or felling, the two species again appear in equal proportions.

Life history

The fruits are very like those of the Roble but produce rather larger seeds. Seedling growth is usually more vigorous than the Roble, and the plants form more persistent and substantial stems more quickly. Annual shoots of 1.5m (5ft) are common in saplings which can attain heights of 7.5m (25ft) in nine years, and girths of 1.8m (6ft) in 30 years. In maturity the Raoul is a smaller tree than the Roble. In the protected woodland of the Malleco National Forest Park, Clarke found the Raoul to average about 21.3m × 0.7m (70ft × 2ft 6in), with straight, bare stems and small, tufted crowns.

Timber

The timber is strong, heavy and close-grained, bright reddish in colour in the heartwood and white to pink in the sapwood, with well-defined growth rings. It seasons easily and gives a smooth finish. In Chile it is one of the best woods for general use, such as furniture, cabinet making, veneers and flooring. Raoul timber grown in Britain seems likely to be inferior to that of Roble.

Introduction, performance and use as an exotic

The Raoul was introduced into England by Elwes in 1902 at the same time as the second and successful introduction of the Roble (although Henry, writing in 1906, claims that the Raoul had not been introduced) and it was planted fairly generally on the landed estates. Records of planting are less numerous than those for the Roble. At Muncaster Castle, Cumbria, a specimen planted in 1923 measured 22.5m × 1.8m (74ft × 5ft 11in) in 1955; Mitchell found it to be 25m × 2.4m (82ft × 8ft 3in) in 1971. Another planted at East Bergholt, Suffolk, in 1915, which, according to Bean, was 13.7m × 0.8m (45ft × 2ft 9in) in 1933, had grown to 25.9m × 2.7m (85ft × 9ft 3in) when Mitchell measured it in 1972. Other recent measurements include the following trees:-

Brodick Castle, Arran, 29.8m × 4m (98ft × 13ft 6in) 1987
Caerhays Castle, Cornwall, 22m × 2.9m (72ft × 9ft 10in) 1988
Frensham Hall, Surrey, 28m × 3.3m (92ft × 10ft 8in) 1987
Minterne, Dorset, 26.8m × 2.4m (88ft × 8ft 2in) 1987
Mt Usher, Co. Wicklow, 32m × 3.3m (105ft × 11ft) 1987

As an ornamental tree, the Raoul is perhaps neater and more compact than the Roble, and with more attractive leaves, but it is more susceptible to frosts and is less happy on poor, silty soils. These two factors somewhat limit its use as a forest tree, but like the Roble it probably has a future as a quick-growing hardwood, although it seems likely to flourish only in the wetter, milder parts of western Britain.

Spanish Chestnut

Castanea sativa

The Spanish or Sweet Chestnut is the largest of our introduced broadleaved trees. Old trees have gigantic boles, remarkable in proportion to their height, clad in deeply longitudinally fissured, spirally angled bark. The trunk supports a domed crown of huge somewhat twisted branches which bear long, narrow, glossy-green leaves with saw-toothed edges.

Distribution and habitat

The Spanish Chestnut is found growing wild in southern Europe, North Africa, Asia Minor, the Caucasus and in north Iran; but it does not appear to extend further east, unlike the walnut, being replaced by other species such as *C. mollissima* in China and *C. crenata* in Japan. Its northern limit in Europe is difficult to define as the original distribution was greatly extended in Roman times. But it seems that this limit runs along the Jura through Switzerland to the Southern Tyrol, and on through Yugoslavia and Hungary to Transylvania. The tree is particularly common in Spain, hence the name, where, in the north, it consorts with the oaks *Q. robur* and *Q. petraea* at elevations between sea level and 760m (2,500ft) and in the south with Cork Oak *Q. suber* in a zone between 820m and 1,640m (2,700 and 5,400ft). In Italy it occurs throughout the Apennines between 910 and 1,220m (3,000 and 4,000ft) and in Sicily on Mt Etna there were, at the turn of the century, some immense and ancient specimens. East of the Adriatic, this tree is a component of the oak and beech forests and in some places extensive woods of it have an understorey of Hazel *Corylus avellana*. It is common in the Greek mountains often forming the lower border of the deciduous forest between 360 and 910m (1,200 and 3,000ft), usually above the region of evergreen shrubs but sometimes descends to sea level. It grows throughout the Caucasus at elevations up to 1,830m (6,000ft) and it extends into north Iran.

Life history

The cord-like 12 to 15cm (5 to 6in) long catkins appear towards the end of June in the axils of the current year leaves. They support both male and female flowers, the former, yellow and sickly smelling, covering most of the catkin, and the latter, much fewer in number, clustered at the base. These develop into prickly fruits, usually containing two round nuts (marrons). The tree starts to fruit at between 20 and 30 years. Growth in early years is rapid, slowing up later, but eventually the trees reach great sizes with heights, when close grown, exceeding 30.5m (100ft) and girths often much in excess of 6.1m (20ft) during very long lives. For instance, Kircher, travelling in Sicily in 1670, saw a colossal hollow chestnut, so big that a whole flock of sheep were penned in it. And many years later Houel found the same tree with a complete family living in its interior. But when Brydone came upon it in 1770 (it was then called The Chestnut of a Hundred Horses) he found it to be five trees growing together to form this huge hollow bole which he measured as 62m (204ft) in circumference. When Druce, from Oxford, saw it in 1900 three of the five trees were still alive, about 21.3m (70ft) tall and still bearing fruit. Elwes mentions a chestnut in Madeira, a planted specimen, which was 49m (160ft) tall and girthed 11.7m (38ft 8in). It was burnt down in 1903.

Timber

Chestnut timber, which closely resembles that of the English Oak, but without the 'silver grain', is hard and heavy. With a naturally lasting heartwood and only a thin layer of sapwood, it is more durable than oak. But it has two serious faults – the grain is often spiral and when grown on sandy soils it develops circular or radial cracks (shakes) which cause it to fly to pieces on the sawbench. Despite this, in former times the timber was in great demand for joists, window frames and general building work.

The tree coppices readily and the resulting poles were much used in the hop gardens. The poles when cleaved and wired together make an attractive type of fencing. Cleaved chestnut palings also make an attractive fencing.

Introduction, performance and use as an exotic

Even as late as 1838, when Loudon was writing, the Sweet Chestnut was considered to have been brought to Europe from Asia Minor by the Greeks around 500 BC; but the extent and age of the forests of this tree in Spain, Italy and the Balkans make this proposition most unlikely. That it was introduced into Britain has not really been disputed except that, in 1885, H. N. Ridley reported the find of charcoal, supposed to be that of chestnut, in a Kent claypit together with palaeolithic implements and the bones of a rhinoceros. But there is no corroborating evidence that this tree was native to Britain in prehistoric times and, as Elwes says 'the specimen may be capable of some other explanation'.

Reason suggests that the Spanish Chestnut was introduced by the Romans who had long valued its fruit in their own country. One of the oldest of the species in England is the tree at Tortworth in Gloucestershire, which Loudon suggests may well have been planted by the Romans. It was a famous tree remarkable for its magnitude in King Stephen's reign (1135), being '197 yards in compass'. It is still there, although somewhat of a wreck, but from its huge, rough, knobbly bole (currently girthing 12.2m (36ft 5in) sprout two great branches, trees in themselves, together with an assortment of similar 'trees' which appear to be layers from the original. It has a strange air of antiquity about it, for it could well be the oldest tree in England. Apart from this one, Loudon mentions many others. The specimen at Bletchworth Park, Surrey, which in the mid-1980's measured 14.9m × 8.6m (49ft × 28ft 5in), is probably one of those which he recorded there as 'some remarkably large chestnuts' in 1838. Similarly, a tree at Nettlecombe Park, Somerset, said by Loudon to be 120 years old in 1837, is almost certainly the same tree measured by Mitchell in 1982 as 24.4m × 8.3m (80ft × 27ft 3in). Elwes reported that a chestnut at Castle Leod, Ross-shire, planted in 1550, measured 23.1m × 6.5m (76ft × 21ft 4in) in 1906. In 1980 it measured 26.8m × 7.8m (88ft × 25ft 7in) – remarkable size and longevity for one so far north. The famous tree at Rossanagh, Co. Wicklow, which was planted in 1718, measured 24.4m × 9m (80ft × 29ft 6in) when Elwes photographed it in 1903. By 1980 it had gone back in height to 21m (69ft) but the girth had increased to 10.4m (34ft 5in).

Other trees recently measured by Mitchell and the Tree Register include:–

Bushey Park, Middlesex, 11m × 9.1m (36ft × 30ft) (hulk with burrs) 1983
Canford School, Dorset, 17m × 13.2m (56ft × 44ft 3in) 1988
Chatsworth, Derbys., 29.8m × 18.7m (98ft × 18ft 6in) (clear bole for 12.2m (40ft) 1983
Croft Castle, Herefords., 22.8m × 8.7m (75ft × 28ft 6in); 21.3m × 8.5m (70ft × 28ft) 1988
Hanworth Hall, Norfolk, 25.9m × 8.1m (85ft × 26ft 9in) 1991
Kings Walden, Herts., 19.8m × 8.7m (65ft × 28ft 6in) 1987
Melsor House, Northants., 18.9m × 11m (62ft × 36ft) 1987
Stourhead, Wilts., 12.8m × 8.1m (42ft × 26ft 9in) 1983
Studley Royal, Yorks., 15.8m × 8.0m (52ft × 26ft 5in) 1983

The Spanish Chestnut is a considerable asset to our tree heritage. In particular, the often vast dimensions of old trees and the rounded crowns of shiny leaves make them significant park ornaments. And the tree's readiness to produce coppice shoots with many uses is another important point in its favour. Its fruit, however, is small and rather tasteless when grown in Britain, and no match for the 'marrons' produced in its more southerly natural habitat in Europe.

There are some attractive varieties of this chestnut. One *albo-marginata* first noticed in 1864, has creamy-white margined leaves, and another *asplenifolia* with leaf lobes drawn out into threadlike points, was first seen in 1838. There is a 15.2m (50ft) tall specimen of this variety at Westonbirt Arboretum.

The Spanish Chestnut although at one time grown in the mid-west states in America, is rarely seen in that country because of its susceptibility to chestnut blight – a disease brought from Europe which has virtually wiped out the native species *Castanea dentata.*

Pin Oak

Quercus palustris

The Pin Oak, a tree of poorly drained, low-lying land, is another of the North East American oaks which adds its colour to the autumn scene. Like the Scarlet Oak *Q. coccinea*, it is a moderately sized tree, but with a narrower, more pyramidal crown. The bristle-tipped, lobed, shiny green leaves are even more indented than those of the Scarlet Oak and grow on many level or downward sweeping branches, with dense, fine shoots – hence the name 'pin'. The straight, unforked trunk, which reaches right up into the crown, is covered in thin, scaly bark. (See illustration facing p.176)

Distribution and habitat

The natural range of the Pin Oak is bordered in the north by a line from New England westwards to north Illinois and south Iowa, including the extreme south of Ontario; on the east, south through Pennsylvania to North Carolina, excluding all the Appalachians; in the south through Tennessee to Central Arkansas; and in the west by north-east Oklahoma and east Kansas.

This oak is normally found in situations which endure frequent but not permanent flooding, and is thus nearly always on flat land and never at altitudes higher than 240m (800ft), unlike Red and Scarlet Oaks.

Because of its ability to grow on flooded land, the Pin Oak, in the south of its range, is often found in pure stands, mixed with only a scattering of other species such as Willow Oak *Q. phellos*, Sweet Gum *Liquidambar styraciflua* and Overcup Oak *Q. lyrata*. On drier sites it may have more associates in mixed stands, such species as Red Maple *Acer rubrum*, Black Tupelo *Nyssa sylvatica* and Honey Locust *Gleditsia triacanthos* being amongst the more common.

Life history

The acorns which are more broad than long sit in saucer-like cups and are produced in huge quantities from the female flowers of trees as young as 25 years and take up to 18 months to ripen, falling in November to germinate in the following spring. Dispersal is by birds, animals and water, although perhaps the majority will fall beneath the parent trees to come up in dense mats which are, however, vulnerable to frequent floods.

Growth is rapid – over 21m (70ft) in 40 years, with girths of 1.8m (6ft). The normal top height in maturity is around 27m (90ft), but an occasional 36m (120ft) tall and 4.5m (15ft) in girth can be found. These oaks are not long-lived, 100 to 150 years being about average.

Timber

The timber suffers from an excess of knots caused by the exceptionally persistent small branches, and only the largest trees go to the mills. The timber is sold as Red Oak and is used for the same purposes.

Introduction, performance and use as an exotic

Although the Pin Oak may have been introduced into Europe around 1770, it did not come to Britain until 1800 at the earliest. Loudon saw some seven-year-old trees at Loddiges Nursery in 1837, which were 4.5m (15ft) tall, but he gives no details of any other specimens, presumably because they were all too small.

The tallest tree which Elwes knew of was one in the Great Park at Windsor which he photographed and measured in 1910 as being 29.5m × 3.2m (97ft × 10ft 8in). He reports hearing of another specimen in the Park which was 34.4m × 3.1m (113ft × 10ft 6in). The tree at Kew, which he measured in 1908 as 17.3m × 3m (57ft × 10ft), with a broken top, cannot be the same tree as that measured there by Mitchell in 1984 as 26.8m × 3m (88ft × 10ft 1in).

Other trees recently measured are:-

Hergest Croft, Herefords., 25.6m × 1.8m (84ft × 6ft 3in) 1978
Holland Park, London, 26.8m × 1.9m (88ft × 6ft 9in) 1981
Hyde Park, London, 26m × 3m (85ft × 10ft 2in) 1988
Melbury, Dorset, 21.3m × 2.2m (70ft × 7ft 7in) 1980

The Pin Oak does not do well in the north and is confined to Southern England where there are some good specimens in large gardens and parks. It grows less well than either the Red or Scarlet Oaks, but colours beautifully. It does not seem to have a future in forestry beyond that of amenity planting in damp soils.

The light branching, neat, narrow habit and beautiful autumn colouring of the Pin Oak have made it a popular ornamental tree in the USA, but with age the lower branches tend to become pendulous, so, if used as a street tree, it must be planted well back from the roadway.

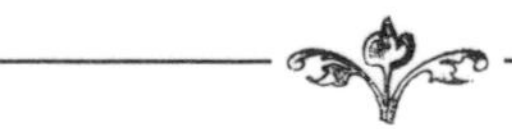

Scarlet Oak

Quercus coccinea

The Scarlet Oak is one of the most ornamental trees of eastern North America; more so than the Red Oak. It is an important contributor to the autumn colour scene with its deeply indented, bristle-tipped, lobed leaves, shiny green in summer, turning a brilliant scarlet in the autumn and remaining on the branches longer than any of the other colourful trees. The crown is narrower and more open than the Red Oak and the tree is altogether less robust. (See illustration facing p.176)

Distribution and habitat

The natural range of the Scarlet Oak includes the whole or part of all states east of the Mississippi River, except Wisconsin and Florida. West of the river it is limited to south-east Missouri. It is also found in Canada, in southern Ontario, west of Toronto.

It is commonly found growing on poor, sandy and gravelly soils and is a typical upland oak established on mountain ridges and higher slopes up to a maximum elevation of 1,520m (5,000ft) in the South Appalachians, although it is normally found at 920m (3,000ft). It reaches its greatest size in the Ohio Basin.

In the north of its range this oak is associated with such trees as White Pine *Pinus strobus*, Red Oak *Q. rubra* and White Oak *Q. alba*; and further south with Short-leaf Pine *P. echinata*, Black Locust *Robinia pseudoacacia*, Sweet Gum *Liquidambar styraciflua* and hickories *Carya* spp.

Life history

Flowering takes place in April and May, and the acorns (rounder and with deeper cups than those of the Red Oak) take two seasons to ripen, nearly half of them being eaten by birds and squirrels. A great many more are destroyed by insects after falling.

This oak grows rapidly and matures early, maximum height being about 30m (100ft) and the girth 3.6m (12ft).

It is one of the least shade tolerant of oaks but manages to maintain a dominant position in some areas by virtue of its quick growth and its ability to do better on poor soils than any of its competitors.

Timber

The timber is almost indistinguishable from that of Red Oak, although slightly inferior in most respects. It is sold as Red Oak and used for the same purposes.

Introduction, performance and use as an exotic

The Scarlet Oak was discovered by Banister and was included in his 1680 catalogue. It was probably he who sent one to his patron, Bishop Compton, as there was a specimen growing in the Bishop's Fulham garden in 1691.

The finest tree which Loudon knew of was at Syon House, Middlesex, measuring 23.4m × 0.8m (77ft × 2ft 9in) in 1838. This tree was no longer there when Elwes looked for it at the turn of the century, and he thought that it and many other specimens of the Scarlet Oak mentioned by Loudon were in fact Red Oaks.

The finest specimen seen by Elwes was at Arley Castle which, in 1904, measured 23.7m × 1.8m (78ft × 6ft 3in). Another at Terling Place, Essex, raised from an acorn planted in 1885, was 7.6m (25ft) tall in 1904, and according to the Forestry Commission 15m (50ft) in 1953. Elwes also noted that where he found Scarlet Oaks and Red Oaks growing together and planted at the same time, the Red Oaks were usually bigger.

The tallest specimen recorded by Mitchell is one at Westonbirt, mentioned but not measured by Elwes, which was 24.4m × 1.4m (80ft × 4ft 10in) in 1956. Twenty-five years later it measured 29m × 1.9m (95ft × 6ft 6in), equal in height to a tree at Elvetham Hall, Hants., which in 1977, was 29m × 2.4m (95ft × 8ft 3in). Other trees more recently measured include the following:-

Bishopthorpe, York, 26m × 3.1m (85ft × 10ft 7in) 1989
Bramscott Hall, Liphook, 25m × 3.1m (82ft × 10ft 6in) 1983
Kippencross House, Perth, 26.8m × 2.4m (88ft × 8ft) 1986
Knaphill Nursery, Surrey, (Splendens), 23.7m × 2.6m (78ft × 8ft 10in) 1989
Saltram House, Devon, 26m × 2.4m (85ft × 8ft 3in) 1984
Wakehurst Place, Sussex, 26.8m × 2.3m (88ft × 7ft 8in) 1988

The Scarlet Oak is generally a slower grower than the Red Oak, but will flourish on poorer soils. It is planted for colour along roadsides and the edges of plantations, but less frequently than Red Oak.

The cultivar 'Splendens' has larger leaves than the type and seems to be a more reliable autumn colourer; more reliable perhaps than the nurseryman's promise for many normal trees which are given this label.

Although the Scarlet Oak is more difficult to move than most trees, and is thus less favoured in the USA than other red oaks for town planting, it has many advantages if special care is taken over the operation. It grows naturally in nearly one-third of the country and is very hardy, as well as having lustrous leaves, brilliant autumn colours and a loosely open habit.

Red Oak

Quercus borealis (rubra)

The Red Oak is one of the largest deciduous trees found growing in eastern North America, and it forms a conspicuous part of the landscape, with its broad, symmetrical crown of dark green, bristle-tipped leaves. The dark grey-to-reddish-brown bole often splits into several stout, upward-growing branches at about 4.5m (15ft) from the ground. The leaves turn a deep red or orange in the autumn and remain on the tree until early winter. (See illustration facing p.176)

Distribution and habitat

The Red Oak has a wide natural range in Canada and the USA, bounded in the north by a line from Nova Scotia to Ontario and Minnesota, south to Louisiana and east to the Atlantic Coast in North Carolina.

It reaches its greatest sizes in the Ohio Valley and on the western slopes of the Alleghennies, usually on north and east aspects, and on lower and middle slopes. In the north of the range it is rarely found at altitudes in excess of 150m (500ft), but in the south it may go up to 1,520m (5,000ft). This oak is usually a component of mixed forest, pure stands being very rare. In the north it is found growing with such trees as Red Pine *Pinus resinosa*, Red Maple *Acer rubrum*, White Pine *P. strobus*, and Eastern Hemlock *Tsuga canadensis*; and in the south with Loblolly Pine *P. elliottii*, Tulip Tree *Liriodendron tulipifera*, Beach *Fagus grandiflora* and Black Locust *Robinia pseudoacacia*.

Life history

The male and female flowers of the Red Oak appear before the leaves, mostly on trees at least 50 years old, the females in the axils of the new leaves producing rather flattened acorns in saucer-like cups. These take two years to mature and after falling in the autumn need a winter in the soil to break dormancy. But the vast proportion is eaten by birds and animals and the population of these oaks is maintained by exceptionally heavy crops of acorns every seven years or so.

The seedlings of this, the most rapid grower of all oaks, are often 2 to 3m (7 to 10ft) tall in five years. The sizes of mature trees vary with site and soil 21 to 27m (70 to 90ft) in height and 1.8 to 2.8m (6 to 9ft) in girth being about average; but in the Ohio Valley trees of up to 49m (160ft) tall and 4.5m (15ft) in girth can be found.

The Red Oak is fairly shade tolerant: more so than ashes, cherries and tulip trees, but less so than hemlocks, maples and hickories. Its ability to spread its crown laterally at the expense of the vertical growth as the tree increases in size, enables it to overtop and shade out competitors.

Timber

The pink to reddish-brown timber is coarse in texture and less attractive than that of the White Oaks, and is also less durable and more likely to warp and check. It is used for flooring, interior joinery, furniture and veneers.

Introduction, performance and use as an exotic

The Red Oak was introduced into France about 1740, and is known to have been grown by Miller in England in 1739. In Loudon's time it was more generally planted than any other American oak, and he records a number of specimens in various parts of the country which were, on

Pin Oak, Lickey Hills, Birmingham (p.172)

Scarlet Oak, Levi Jackson Park, Kentucky (p.173)

Red Oak, Levi Jackson Park, Kentucky (p.175)

Holm Oak, Howick Hall, Northumberland (p.179)

White Mulberry, ripe fruits. (p.184)

Katsura Tree, Westonbirt, Glos. (p.186)

Cucumber Tree, Tennessee Valley Authority Park (p.188)

Southern Magnolia, Orlando, Florida (p.189)

average, between 12 and 15m (40 and 50ft) tall in 1838, although one at Stratfield Saye was 30m (100ft) tall and the tallest he had encountered. But this oak does not seem to be long-lived in Britain. Elwes did not find any of Loudon's trees and the tallest and largest measured by Mitchell were not known to Elwes. He photographed one at Kedleston, Derbyshire, which, in 1903, was 27.4m × 3.3m (90ft × 11ft), with a main trunk of 9m (30ft); another at Pains Hill, Surrey, although only 24m (80ft) tall, was 5.7m (19ft) in girth. A third tree which he lists, at Whitton, Durham, measuring 27.4m × 2.3m (90ft × 7ft 10in) in 1903 was 150 years old; probably getting on for the life span of the Red Oak in Britain.

Recent measurements include the following trees:-

Forde Abbey, Dorset, 20m × 4m (66ft × 13ft 6in) 1988
Glanrhos, Powys, 26m × 3.8m (85ft × 12ft 10in) 1988
Gliffaes, Powys, 20m × 4.8m (66ft × 16ft) 1984
Gray House, Dundee, 22m × 4.3m (72ft × 14ft 8in) 1986
Southampton Common Cemetery, 21.3m × 4.1m (70ft × 13ft 10in) 1982
The Whittern, Herefords., 24.4m × 4.5m (80ft × 15ft 1in) 1987

The Red Oak is a very fast grower, for a hardwood, in Britain – shoots of up to 1.8m (6ft) long in a year being common on good sites; and it will grow much faster than our native oaks on poor soils, at least in the early years. So it could have a place as a producer of a good general purpose hardwood but one which could in no way replace that of our native species for specialised uses such as joinery.

It is mainly planted as a forest tree along the margins of conifer plantations and on sites of particular scenic value where autumn colours are seen to advantage.

In the USA, this native oak is more often planted for amenity purposes than any other oak. It transplants easily, unlike most of the family, and grows vigorously, as well as withstanding city pollution. It is particularly favoured for avenue planting.

Japanese Chestnut Oak

Quercus acutissima

The Japanese Chestnut Oak or Sawtooth Oak is a medium sized, deciduous tree, up to 21m (70ft) tall, with a bole that has dark grey bark, deeply irregular and longitudinally fissured. The trunk bears an open, rounded crown of oblong, saw-edged leaves, deep green above and pale green beneath.

Distribution and habitat

The natural range of this oak stretches right across eastern Asia from the Himalayas in the west to Korea and Japan in the east. In China it is one of the many components with other oaks such as the very similar *Q. variabilis* and *Q. dentata*, of the deciduous forests which cover the lower slopes of the mountains below 1,520m (5,000ft) in the central and eastern provinces from Szechuan to Kiangsu. In Japan, its absence from remote mountain areas and its abundance near villages cast doubt upon its status as an indigenous species. However, it is found throughout the warm temperate zone from central Honshu south to Kyushu. It has long been planted for fuel.

Life history

The flowers appear with the leaves in April and May, the males in pendant catkins and the females singly or in pairs on the upper parts of the twigs. These develop into acorns which ripen in the autumn of the following year, as shiny brown nuts set in cups bearing numerous recurved scales.

Timber

The hard, strong timber is used for building, turnery, house utensils and especially for charcoal.

Introduction, performance and use as an exotic

The Japanese Chestnut Oak was introduced into Britain in 1862, but is very rarely seen. It is not hardy except in some places in southern England and in Ireland where it can be found in a few large gardens. It is never of any great size.

Recent measurements include the following trees:-

Higham Court, Glos., 22.8m × 2.8m (75ft × 7ft 8in) 1988
Royal Botanic Gardens, Kew, 21.3m × 1.6m (70ft × 5ft 6in) 1987
Tregrehan, Cornwall, 17m × 1.6m (56ft × 5ft 8in) 1987

In America, where this tree was introduced around 1880 by acorns sent to the Arnold Arboretum from Peking, it has fared better becoming naturalised in many places. Its relatively rapid growth, neat rounded crown and late retention of leaves which turn an attractive yellow in the autumn, have made it increasingly popular as an ornamental, especially in the eastern States where, surprisingly, it has withstood winter temperatures of -29C (-20F).

Holm Oak

Quercus ilex

The Holm Oak, a native of the Mediterranean area, is the largest non-coniferous tree grown in Britain. The leaves which are often variable in shape, form a domed crown to the tree. They emerge silvery-white and woolly, turning shiny green above and brown beneath, and growing on slender, ascending branches. These branches are supported on a considerable trunk clad in brown to black bark which is lightly cracked into small, square, scaly plates. (See illustration facing p.176)

Distribution and habitat

The Holm Oak has a wide distribution in the Mediterranean area, being found in Spain, Portugal, France, Italy, Dalmatia, Greece and Syria, as well as in Morocco, Algiers and Tunis. It grows at sea level, where it withstands salt winds better than any other oak, and up to a height of 1,520m (5,000ft), although there it may be no more than a scrubby bush.

At low altitudes in France, it is seldom found in pure stands but grows in association with Aleppo Pine *Pinus halepensis* and higher up, at around 760m (2,500ft), with Downy Oak *Quercus pubescens*. Further south, in Italy, Stone Pine *Pinus pinea* and Maritime Pine *Pinus pinaster* are associated species. In Portugal, it is often found growing with Cork Oak *Quercus suber* with which it frequently hybridises.

Although generally a tree of relatively mild climates, it can bear a considerable degree of winter cold. Elwes reported that a tree at Grignon in France survived a temperature of -23C (-10F) in 1871 uninjured. Severe winters usually turn the leaves a pale brown without affecting the trees' welfare.

Life history

The male catkins appear in spring from the axils of the leaves, the gold stamens shining against a grey and dark green leafy background. The female catkins open a week or two later on short peduncles, developing into rather variably shaped acorns, often half-enclosed in their cups. Early growth of the seedlings, which emerge the following spring, is slow largely because much of their energy is spent in putting down an exceptionally long tap root. Later, growth speeds up and a sapling may be as tall as 4.5m (15ft) after 10 years, but as the tree approaches its half century, height growth tends to give way to crown spread.

It is probable that Holm Oaks live to a great age but it seems doubtful if Pliny's contention was correct that some trees he knew were as much as

1,500 years old. Slight scepticism must also be shown of his claim that a tree in the Vatican during his time was older than Rome itself. Nevertheless, other ancients also wrote of Holm Oaks as extremely old and sacred trees, so their longevity cannot be doubted.

Timber

The dark brown timber is hard, heavy and durable but tends to warp and split on seasoning and is difficult to work. In any case, it is not available in sufficient quantity in Britain to be of any commercial value. In southern Europe it was formerly used for ship building, cart wheels, tool handles and other items where a hard and shock-resisting timber was required. It also makes excellent charcoal.

Introduction, performance and use as an exotic

The first mention of Holm Oaks in Britain is by Gerard, writing in 1597, when he called it 'a stranger in England'. But Johnson, in his edition of Gerard, published in 1636, says that, in 1581, Clusius spoke of three trees in London, one of which bore acorns. Parkinson mentioned the same tree in 1640 as being 'in the King's privie garden in Whitehall'. By the time Evelyn saw it in 1678 it had become a 'sickly imp'.

Loudon, writing in 1838, gives details of many specimens but only two of them seem to have been found by Elwes. One at Wilton House, Wilts., Loudon says was 3m (10ft) in girth in 1810; when Elwes measured it a century later it girthed 5.4m (18ft) and had a spread of branches 90 paces in circumference. Another specimen at Stackpole Court, Pembs., noted by Loudon as measuring 23.7m × 2.2m (78ft × 7ft 6in) in 1838, and aged 100 years, was found by Elwes, in 1906, to be 24.4m × 2.8m (80ft × 9ft 8in), although split to the ground and kept together by a chain. Another at the same place, measured by Elwes was 18.3m × 4.2m (60ft × 14ft 5in) in 1906. Mitchell also found this tree in 1984; its height remained the same but the girth had increased to 5m (17ft 2in). A tree at Frogmore, Berks., which Elwes says was raised from an Algerian acorn and was 15.2m × 4.6m (50ft × 15ft 6in) in 1904, Mitchell found to be the same height but girthing 5.4m (17ft 11in) in 1982. A specimen at Melbury House, Dorset, which Loudon recorded as being 40 years old in 1838 and 9m × 0.9m (30ft × 3ft) may be the same tree which Mitchell measured as 26.8m × 4.2m (88ft × 14ft 4in) in 1989. Other trees which he has recently measured are:-

Biel, E. Lothian, 26m × 3.7m (85ft × 12ft 8in) 1985
Chilham Castle, Kent (planted, 1619 ?), 24.4m × 7.4m (80ft × 24ft 7in) (at 30cm (1ft)) 1983
Glendurgan, Cornwall, 24.4m × 4.8m (80ft × 16ft 4in) 1984
Henham Hall, Suffolk, 25.6m × 4.1m (84ft × 13ft 9in) 1976
Howick Hall, Northumberland, 22.8m × 6.5m (75ft × 21ft 5in) 1978
Knaphill Nursery, Surrey, 22m × 3.8m (72ft × 12ft 8in) 1987

Tregothnan, Cornwall, 26.8m × 5.1m (88ft × 17ft 3in) 1989
Woodstock, Co. Kilkenny, 29m × 4m (95ft × 13ft 6in) 1989

The Holm Oak has been planted extensively throughout Britain and thrives in parks and gardens everywhere except in northern Scotland. Its resistance to salt-laden winds makes it abundant in coastal areas and particularly so in north Norfolk on the Earl of Leicester's Holkham Estate where there is a fine avenue and groves of this tree.

In the USA, where it was introduced in colonial times, the Holm oak has proved itself to be one of the best evergreen oaks for coastal planting, both for shelter and for shade. It is found mainly in the south eastern and north western coastal areas of the country.

Chinese Elm

Ulmus parvifolia

The Chinese Elm is a medium-sized, semi-evergreen tree, up to 15m (50ft) tall, with a slender trunk clad in grey, scaly bark, peeling off in roundish flakes to expose brown under-bark, and bearing numerous spreading and upward-sweeping branches. These form a dense, rounded crown of small, dark green, smooth, shiny leaves which persist on the tree until January or, in some cases, until new leaves appear in spring.

Distribution and habitat

The Chinese Elm is a native of central and south-east China, on lower slopes of the mountains bordering the Yangtse Valley, and of Taiwan and Japan. In Shensi Province of China it occurs as a minor component of the oak forests on the lower slopes of the mountains below 1,220m (4,000ft) with such trees as the Pagoda Tree *Sophora japonica*, Chinese Thuya *Thuya orientalis*, Walnut *Juglans regia* and Chinese Juniper *Juniperus chinensis*; and further south, in Kiangsu, with Sweet Gum *Liquidambar formosana*, Prickly Castor-oil Tree *Kalopanax pictus* and David's Maple *Acer davidii*.

In Japan it is confined to the southern parts of Honshu and to Shikoku and Kyushu where it is rarely more than a large shrub 4.5 to 6m (15 to 20ft) tall.

Life history

Unlike most other elms, the flowers of the Chinese Elm appear in late summer or autumn in the axils of the leaves which have been on the tree since the spring. The female flowers produce nutlets set in the centre of a circular wing and these are dispersed by wind.

Mature trees rarely reach more than 18m (60ft) in height; about 12m (40ft), with girths of 1.2m (4ft) or so is more usual.

Timber

The timber is hard, heavy and tough. However, trunks large enough to be turned into planks are seldom found so it is of little commercial value.

Introduction, performance and use as an exotic

The Chinese Elm first came to Europe when Abbe Gallois brought it to France in the reign of Louis XV, supposing it to be the tea plant. James Main, an Essex plant collector, was the first to bring plants home to Britain from China in 1794.

Elwes noted how rare it was in cultivation at the turn of the century. A tree at Kew died in 1912, when it measured 10.6m × 0.9m (35ft × 3ft). It used to produce flowers in November but never set fruit. Another specimen at Beauport, Sussex, grafted onto a Common Elm, was 12.2m × 1.1m (40ft × 3ft 8in) in 1911. This tree is still very rarely planted in Britain. A forest plot, established in Kent in 1934, failed.

Recent measurements include the following trees:-

Dyffryn, Glam., 14m × 1.1m (46ft × 3ft 9in) 1984
Stanmer Park, Sussex, 15.8m × 0.9m (52ft × 3ft 2in) 1988

In the USA, where it was introduced by Thomas Hogg in 1865, it is highly regarded wherever it can be grown (it is not hardy in the north), especially in California where near-evergreen forms have been selected and developed. It is resistant to elm disease and is much less prone to breakage than the other fairly commonly planted Siberian Elm *Ulmus pumila* with which it is often confused. A cultivar 'Pendens', remarkable for its wide-spreading branches, which are frequently pendulous and touch the ground, is particularly popular. In some individuals the bark peels to reveal a lighter colour beneath in roundish patches, which greatly adds to their attraction.

Keaki

Zelkova serrata

The Keaki is a large, elm-like, deciduous tree native to Japan and China. The bole is grey and rather smooth, like the Beech, and the branches have a tendency to sweep upwards, end in very slender, straight twigs. These bear the leaves which are quite small, pointed and sharply-toothed.

Distribution and habitat

The natural range of the Keaki extends throughout most of Japan except Hokkaido in the north and the extreme southern part of Kyushu in the south. It also extends into Korea and to parts of Central China from Shensi eastwards to Chekiang. Although it prefers low-lying areas, with deep, permeable soil, especially along streams, it does grow in the Japanese mountains up to an altitude of 1,220m (4,000ft).

In Japan it is normally found growing with other hardwoods such as Siebold's Beech *Fagus crenata*, Oak *Quercus crispula* and various maples *Acer* spp.; and in China with Chinese Fir *Cunninghamia lanceolata*, Sweet Gum *Liquidambar formosana* and Weeping Cypress *Cupressus funebris*.

Life history

The flowers appear in April and May, at the same time as the leaves, the males in the axils of the lower leaves and the females in the axils of the upper leaves. These develop into little, nut-like fruits with ridges on the surface and two persistent styles at the apex.

The shade-tolerant seedlings grow rapidly in the forest, quickly becoming rather open, upswept trees with remarkably straight trunks, which can be 1.5m (5ft) in girth in 60 years, 2.7m (9ft) in 120 years and 3.6m (12ft) in 180 years. A notable forest-grown tree mentioned by Elwes was 37m (123ft) tall and 3.7m (12.5ft) in girth, with the trunk free of branches for 17.3m (57ft). Another felled near Tokyo in 1874 had a girth, at 90cm (3ft) from the ground, of 11.5m (38ft). But the biggest one which Elwes actually saw in Japan was in a grove which may have been planted; it measured 35m × 6m (115ft × 20ft).

The Keaki is a particularly good tree for bonsai work and is a very popular subject for this uniquely Japanese practice.

Timber

The timber of the Keaki is one of the most valuable hardwoods in Japan. It resembles that of Elm, but has the bending properties of Ash, and is stronger even than Oak, and very durable. Many of the oldest temples in Japan are built of this timber which seems to be indestructible under cover. It is also used for construction work, boat building and for furniture, and pieces cut from burrs are especially prized.

Introduction, performance and use as an exotic

The Keaki was introduced into Britain in 1862 by J. Gould Veitch who sent seeds home from Japan; but Siebold seems to have brought it to Europe a few years earlier.

The damage by frost to early plantings caused some pessimism about its future in Britain which was probably justified. Elwes could find none taller than 9m (30ft). He mentions a specimen at Kew which was about this height in the early 1900's. In 1957, according to Maynard Greville, this same tree measured 18.3m × 1.4m (60ft × 4ft 8in). And a specimen at Hergest Croft, Herefords., which he measured as being 15.2m × 1.8m (50ft × 6ft) in 1961, Mitchell found to be 19m × 1.9m (62ft × 6ft 9in) in 1989. Other trees recently measured include the following:-

Battersea Park, London, 18.3m × 2.8m (60ft × 9ft 6in) 1989
Chelsea Physic Garden, 18.3m × 1.6m (60ft × 5ft 6in) 1989
Royal Botanic Garden, Edinburgh, 20m × 1.9m (66ft × 6ft 6in) 1983
Horstead Keynes, Sussex, 19m × 2.7m (62ft × 9ft 4in) 1982
Tilgate Park, Surrey, 18.9m × 2.4m (62ft × 8ft) 1988

The Keaki is not an easy tree to grow in Britain as frost cuts it down to ground level and it then tends to become bushy. But this does not seem to be so in the USA. This tree was introduced there in the same year, 1862, when a Dr Hall planted two trees in his Rhode Island garden. In 31 years they were 15m (50ft) tall, with girths of 0.9m (3ft), and regularly produced flowers and fruit, with hundreds of seedlings growing beneath them. In recent years this tree has come to the fore in the USA as a disease-resistant substitute for the American Elm, especially in the northern States. It has also proved itself to be very accommodating about soil and climate. It is difficult to explain why it does not suffer frost damage there, in weather which is much more severe than that experienced in Britain. Neither does it suffer on the Continent where early plantings in Germany and Denmark grew steadily and unharmed.

White Mulberry

Morus alba

The White Mulberry is a small tree, rarely more than 12m (40ft) tall, with a bole, often tortuous, covered in thick, furrowed bark occasionally disfigured with large burrs. It has a narrow crown of very variable but generally oval leaves with toothed edges. It is these leaves which are the favourite food of silkworms. (See illustration facing p.177)

Distribution and habitat

Like so many trees, particularly those from Asia, which have been cultivated for centuries, the exact natural range of the White Mulberry is rather doubtful, but it certainly occurs wild in the mountains of central and northern China, and probably also in Japan. In Shensi and Kansu Provinces it is a scattered component of the oak forests below 1,220m (4,000ft), together with Pagoda Tree *Sophora japonica*, Walnut *Juglans regia* and Juniper *Juniperus chinensis*. It also occurs in the broadleaved forests of the Upper Yangtse Provinces, as well as in the much harsher conditions prevailing in the oases of the Sinkiang Desert where it is in association with other deciduous broadleaved species such as Black Poplar *Populus nigra* and Siberian Elm *Ulmus pumila*. In Japan it occurs, apparently wild, in the virgin forests of central Honshu and, according to Mayr, on Bonin Island 960 kilometres (600 miles) south of Japan.

Life history

The flowers appear in May in the leaf axils and at the base of the new shoots, the females on stalked, cylindrical spikes. The fruits which are sweet and juicy but insipid, are about 12mm (0.5in) in diameter and resemble pale, pinkish blackberries, each of the aggregated drupes containing a seed.

The seedlings are very tiny and vulnerable but grow fairly quickly to reach a height of 9 to 12m (30 to 40ft) when mature; except those in Japan which are said to be much taller. Indeed, Mayr found one on Bonin Islands which was 30m (100ft) tall. Variations in size are probably part of the tree's generally variable character.

Timber

The Chinese once used the timber of White Mulberry for cabinet work, musical instruments, carving and turnery.

Introduction, performance and use as an exotic

The cultivation of the White Mulberry for feeding silkworms is supposed to have started in China around 2,700 B.C. From China the art of silk production spread to Persia and to Greece after Alexander the Great's expedition in 350 B.C. From there it spread to Rome, with Sicily being found to have suitable conditions for the growing of this tree. In the late 15th century the French started a silk industry and the planting of the Mulberries in the southern parts became quite widespread. But it was not until 1609 that the British took it seriously, when James I became interested, although the White Mulberry had been introduced in 1596. Attempts at growing these trees satisfactory to the needs of the silkworms failed, possibly because the climate did not allow the leaves to develop the correct chemistry for the insects' thread production.

As an ornamental and fruit-producing tree the White Mulberry has never really succeeded in Britain, nor does it attain any great size. Loudon mentions a specimen at Syon House, Middlesex, which was 13.7m (45ft) tall in 1838, but Elwes never found it. The tallest he knew of was one at Beauport, Sussex, which was 7.6m × 1m (25ft × 3ft 3in). Specimens recently measured include the following:-

Cirencester Park, Glos., 8m × 1.6m (26ft × 5ft 6in) 1989
Dean's Yard, London, 12.8m × 0.8m (42ft × 2ft 10in) 1988
Dyffryn, Glam., 8.5m × 1.1m (28ft × 3ft 9in) 1988
Oxford Botanic Gardens, 12.2m × 2.4m (40ft × 7ft 11in) 1987

In southern Europe, where this tree has long been cultivated, it is common, grown for its fruit and for shade. There are some very venerable specimens in towns and cities.

The White Mulberry was introduced into the USA in colonial times and, despite Wyman's opinion that 'none of the mulberries should be considered good for general ornamental planting and all might well be placed on the discard list', it has been widely planted in most States and has even become naturalised in the east of the country. A non-fruiting cultivar 'Kingan' is useful for street planting, and the variety *tartarica* seems to be the hardiest.

This tree flourishes best in warm-temperate and sub-tropical climates. In India, trials for timber production have met with some success and in the Central African Highlands it is grown for fruit production.

Katsura Tree

Cercidiphyllum japonicum

A native of China and Japan, the Katsura tree is one of the largest deciduous trees in those countries. The huge, domed and remarkably symmetrical crown of nearly round, bluish-green leaves, heart-shaped at their bases, is supported by one or more trunks clad in dark greyish-brown, deeply furrowed, bark, separating into flakes. (See illustration facing p.177)

Distribution and habitat

The Katsura Tree is common in Western China, particularly in western Szechwan and western Hupeh. On the Sinkiang-Szechwan border it occurs in the lower limits of the mixed montane forests below 1,070m (3,500ft), in association with such trees as Chinese Red-barked Birch *Betula albo-sinensis* and the Prickly Castor-oil Tree *Kalopanax pictus*, as well as with hemlocks and firs, all with an understorey of bamboos. Further east in more sheltered valleys, where it grows to great sizes on rich, moist, riverine soils, it is a component of the largely evergreen broadleaved forests of oaks and chestnuts, with other deciduous species including the Dove Tree *Davidia involucrata*.

In Japan it is widespread in Hokkaido and in the northern half of Honshu, but becomes more scattered further south, although it is still present down to southern Kyushu.

Life history

The male and female flowers are borne on separate trees and open in May before the leaves appear, the females developing into bunches of four to six claw-like pods on stalks at the base of the leaves. They ripen in October when they shed their winged seeds.

Growth is very rapid and the trees attain great sizes. Wilson reported specimens in China which topped 40m (130ft), with girths of up to 5.4m (18ft). One very old specimen with a broken top had a remarkable girth of 16.7m (55ft). In Japan, too, heights exceeding 30m (100ft) are common.

Timber

The timber is light brown in colour, light in weight, soft and straight-grained, with a marked difference in tone between the sapwood and heartwood. It is highly valued for building work, furniture and cabinet making.

Introduction, performance and use as an exotic

The Katsura Tree was introduced into Britain from Japan in 1881, and was then thought to be indigenous only to that country. Later, in 1910, Wilson discovered it in China.

In Britain this tree rarely reaches any great size; about 18m (60ft) being the average maximum height, and this takes many years because of its extreme susceptibility to late frost damage, which often causes multi-stemmed growth. It is, however, a very handsome subject for a garden in areas where spring frosts are not too prevalent, both because of its symmetrical, rounded shape and its attractive oval to circular leaves which turn a beautiful orange and yellow in the autumn. It is particularly striking when planted together with dark green, fastigiate conifers.

Recent measurements include the following trees:-

Ashbourne, Co. Cork, 21.3m × 1.8m (70ft × 6ft 4in) 1987
Batsford Park, Glos., 22m × 1.2m (72ft × 3ft 10in) 1987
Westonbirt, Glos., (Main), 22m × 1.6m (72ft × 5ft 7in), (Victoria Gld.), 26.8m × 1.3m (88ft × 4ft 6in) 1986
Wisley, Surrey, 15.8m × 1.8m (52ft × 6ft) 1987

According to Wyman, the Katsura Tree was introduced into the USA in 1865 where it has become a popular ornamental both because of its graceful habit and of the resistance of its leaves to insect attack. The multi-stemmed forms serve as shade trees and the more fastigiate single stemmed species for street planting.

Cucumber Tree or Mountain Magnolia

Magnolia acuminata

The Cucumber Tree is the largest of the seven magnolias native to North America, and the only one with rough bark, greyish-brown in colour and divided into narrow ridges and vertical grooves. The distinctive bole, 15 to 24m (50 to 80ft) tall, supports a conic crown, with generally upswept branches, but with the lower ones sweeping down to the ground. The leaves are thin, 25cm (10in) long, oblong and sharp-pointed. The bell-shaped greenish flowers are followed by fleshy 12cm (7in) long cucumber-like fruits. (See illustration facing p.177)

Distribution and habitat

The natural range of the Cucumber Tree extends from southern Ontario to central New York, through south eastern Illinois and south along the Appalachians to southern Alabama and central Mississippi; and also in Arkansas and southern Missouri. It is never very numerous, especially in the north, and is usually scattered singly through the forest. It is largely an upland tree sometimes growing as high up as 1,220m (4,000ft) in the south, but reaching its maximum development on moist fertile soils on the low slopes at the base of the Smoky Mountains in the Carolinas and Tennessee.

In the north it is found growing sparsely amongst such trees as the Sugar Maple *Acer sachharum*, White Oak *Quercus bicolor* and White Ash *Fraxinus americana*; and in the south with the Tulip Tree *Liriodendron tulipifera* and Hickories *Carya* spp.

Life history

The fleshy fruits, green at first, then pink and finally purplish red, bear scarlet seeds scattered on their surfaces, like corn on the cob. When ripe, these seeds hang from the fruit on slender white threads before falling to germinate in the following spring or sometimes in the spring following that.The seedlings are so intolerant of shade that few survive; but those that do, on favourable sites, grow rapidly at first reaching 24m (80ft) in height in some 120 years.

This tree is a popular ornamental in the Eastern States because of its attractive symmetry of form and its almost tropical foliage.

Timber

The yellow-brown timber is soft, durable, close-grained and easily worked. It is used for cheap furniture, cabinet work, crates and flooring.

Introduction, performance and use as an exotic

The Mountain Magnolia was discovered by one, John Bertram, who sent seeds and plants to Peter Collinson in 1736. As it was easily propagated by layers it soon became well known in gardens in Britain and on the Continent. Trees grown from the seed sown by Collinson flowered for the first time 26 years later, in 1762.

Loudon gives details of a number of specimens of which the largest was at Thorndon Hall, Essex, which girthed over 2m (7ft); and another at Syon House, Middlesex, which was 15m (49ft) tall in 1838, and could be the same tree as that measured by A.B. Jackson in 1908 as 15.5m × 1m (51ft × 3ft 7in).

Elwes photographed a specimen at West Dean, Sussex, which measured 18.3m × 2m (60ft × 7ft) in 1906; Mitchell found it to be 18m × 2.8m (59ft × 9ft 6in) in 1974. A tree at Fawley Court, Henley, Oxon. which Elwes measured as 12.2m × 1.6m (40ft × 5ft 6in) in 1907, girthed 2.2m (7ft 8in) according to Gardner who saw it in 1950; and Mitchell recorded in 1985 that it had grown to 18.3m × 3m (60ft × 9ft 10in).

Other trees measured include:-

Brockenhurst Park, Hants., 21.3m × 2.9m (70ft × 9ft 9in) 1986
Claremont House, Surrey, 19m × 2.1m (62ft × 7ft 4in) 1986
Margam Park, Glam., 26.8m × 2m (88ft × 7ft 2in) 1985
Portmeirion, Gwynedd, 22.8m × 1.8m (75ft × 6ft 4in) 1987
Westonbirt, Glos., 26m × 1.9m (85ft × 6ft 8in) 1988

In Britain this magnolia is generally hardy and, although in no way competing in flowering beauty with some of the Asiatic species, its quick growth, ultimate large size and neat, symmetrical habit, make it a useful ornamental tree for large gardens and parks. It is also used as a root stock for the grafting of other more showy species of magnolia.

In the USA, this native species is rated an excellent ornamental tree for its foliage and habit. It is often planted where space allows the fast-growing, spreading branches to develop.

Southern Magnolia

Magnolia grandiflora

The Southern Magnolia is a tree of the mixed forests of the south eastern states of the USA, and one of the most magnificent flowering trees of that country. The stout bole with its brownish, scaly bark is often free of branches for up to 12m (40ft). It supports a conic crown of lustrous, leathery, evergreen leaves, and in spring, large creamy white flowers at a height of 24m (80ft). (See illustration facing p.177)

Distribution and habitat

The Southern Magnolia enjoys a warm temperate or semi-tropical climate and is found in a 160 kilometres (100 mile) strip along the Atlantic coast from North Carolina to Florida and west along the Gulf to south east Texas.

It grows best on moist, well-drained soils along streams and near swamps in the coastal plain and is nowhere found at altitudes higher than 150m (500ft). It is rarely found in pure stands but scattered through the forest in company with Cherry Oak *Quercus falcata*, Sweet Gum *Liquidambar styraciflua*, Tulip Tree *Liriodendron tulipifera*, Black Tupelo *Nyssa sylvatica* and Shumard Oak *Q. shumardii*.

Life history

Southern Magnolias have been known to produce their 15cm (6in) wide flowers when as young as 10 years old. When more mature, the flowering is particularly prolific, each new shoot on every limb carrying a flower which develops into a fruit shaped like a hen's egg. As the fruit ripens it bears many scarlet seeds on its surface which are dispersed by birds and animals, such as quails, turkeys, squirrels and possums. Germination beneath the parent tree is sparse, tending to suggest that the seeds fare best after passing through the alimentary system of birds and beasts.

Seedlings are very frost tender, which probably accounts for this tree's fairly limited distribution; but once they get going growth is rapid – anything between 45 and 60cm (1.5 and 2ft) in a year. Eventual heights reached are 18 to 24m (60 to 80ft) and girths of 1.8 to 2.8m (6 to 9ft) at ages between 80 and 100 years. The best growth is attained in Florida where heights of 38m (125ft) have been recorded. The largest specimen so far measured had a trunk girthing 5.4m (18ft) at the base.

Young trees are good shade-bearers, but they need plenty of sun later in life.

Timber

The yellowish timber has a close, uniform texture with a straight grain. It is not very strong or durable but withstands shock well. It is used for making furniture and for Venetian blinds.

Introduction, performance and use as an exotic

The Southern Magnolia's date of introduction into Britain is uncertain. Sir John Colliton had a specimen growing in his Exmouth, Devon, garden before 1737, which may well have been the first one brought to England, if not to Europe. It was felled by a mistake in 1794, but even then was only 1.5m (5ft) tall, largely because of the mutilation caused by the 'layering' exploits of the many gardeners who came from all over the country.

Loudon reported that in 1838 there were many specimens, especially in Devon, taken by this layering, up to 9m (30ft) tall, most of them growing against walls, but some were standard trees. One such at Powderham Castle was growing well as a standard of over 9m (30ft) tall in 1838, although it appears to have grown little more when Elwes saw it some 60 years later.

Although this magnolia can survive temperatures down to freezing, it grows very slowly in areas where the mean temperatures are low. In fact, it fares best, even in mild areas, when planted against a wall and is most commonly found in such a situation. Free standing trees are only occasionally to be seen, mostly in Southern England and Ireland. Two of the best known to Mitchell are one at Goodwood, Sussex, which measured 11m × 1.2m (36ft × 4ft 3in) in 1980, and the other at Cockington Court, Devon, which, in 1978, was 10.3m × 1.2m (34ft × 4ft 3in). Everywhere, whatever form it takes, it tends to flower very late in the year.

Among the best are:-

Ashbourne House, Co. Cork, 10m × 1.5m (33ft × 5ft 5in) 1987
Aston Court, Avon, 10m × 1.5m (33ft × 5ft) 1985
Fota, Co. Cork, 12.2m × 1.3m (40ft × 4ft 6in) 1984
Nonsuch Park, Epsom, 10m × 2.1m (33ft × 7ft 5in); 11m × 1.8m (36ft × 6ft 4in) 1989

A cultivar 'Exmouth', which starts to flower at an earlier age than the type, is probably derived from Sir John Colliton's original tree. Another cultivar 'Goliath' which bears flowers up to 30cm (1ft) in diameter and has broader leaves than the type, was first distributed by Caledonian Nurseries, Guernsey in 1910.

Generally speaking, this magnolia grows to greater dimensions on the Continent. Elwes reported specimens up to 21m (70ft) tall in Italy and 18m (60ft) in France; and even in the early 19th century Loudon found them to be 15 to 18m (50 to 60ft) tall in those countries.

In the USA this magnificent native magnolia is extensively planted in southern areas and even as far north as Pennsylvania, and on the west coast right up to British Columbia. The cultivar 'Exmouth', although it does not bloom at such an early age as the type in that country is popular because of its pyramidal habit and its narrow leaves with rusty undersides. 'Goliath' is also widely planted.

Campbell's Magnolia

Magnolia campbellii

In overall dimensions and size of flower, as well as flower colour, Campbell's Magnolia is the doyen of the genus. The domed crown can be up to 30m (100ft) tall in the Himalayan forests. The long, upturned branches bearing oval, shiny-green leaves are supported by a slightly sinuous but stout, smooth, dark grey bole. For a short period in spring this Magnolia is covered with large, rosy coloured flowers. (See illustration facing p.192.)

Distribution and habitat

The natural range of Campbell's Magnolia is in the eastern Himalayas, stretching from eastern Nepal through Sikkim to Bhutan where it is found as a very minor component of the wet temperate forests of those regions at altitudes between 2,140 and 3,050m (7,000 and 10,000ft) on rich, porous soils of the steep mountainsides. Although forming only about two per cent of the tree species in these forests, these magnolias become very conspicuous as they burst into flower before leafing and show up to perfection in the relatively drab canopy of the other trees, such as the Oaks *Quercus lamellosa, Q. lineata,* Chestnuts *Castanopsis tribuloides* and Maples *Acer campbelli.*

Life history

The large, sweet-scented, pink, cup-shaped flowers 15 to 25cm (6 to 10in) across, appear in April, before the leaves. The thick, fleshy petals fall in a pink carpet round the tree in a matter of days. The fruit is 15cm (6in) long, sausage-shaped, and studded with red seeds.

As with so many of the Himalayan trees, the largest specimens of this magnolia were seen only in the early days of the Raj, before exploitation decimated them. According to Henry, trees of up to 45m (150ft) tall and anything up to 6m (20ft) in girth were common in Sikkim in the middle of the last century. Now such specimens are never seen, for they were in great demand for building timber and for tea-box boards for the expanding tea industry of those areas. Nowadays, although these magnolias, drawn up in the dense forest, may be up to 30m (100ft) tall, they rarely exceed 2.4 to 3m (8 to 10ft) in girth.

Timber

The timber is rarely cut for commercial purposes.

Introduction, performance and use as an exotic

The introduction of Campbell's Magnolia into Britain by means of seeds

Campbell's Magnolia, Trewithen, Cornwall (p.192)

Tulip Tree Forest, Levi Jackson Park, Kentucky (p.195)

Saucer Magnolia, Batsford Park, Glos. (p.193)

Sweet Gum, Smoky Mts National Park, Kentucky (p.197)

Oriental Plane, Troodos Mts, Cyprus (p.200)

Black Cherry, Pijven, Belgium (p.202)

Honey Locust, Selly Oak, Birmingham (p.204)

in the mid-19th century proved difficult, as the seeds decayed in transit. Eventually, live plants were sent from Calcutta Botanic Gardens to Kew about 1880, according to Elwes, but he goes on to say that a specimen flowered in a garden in Cork, Ireland, in 1885, so it was probably introduced earlier than 1880. A tree at Veitch's Nursery in Exeter flowered in 1898, when 12 years old, and another in Cork produced 147 flowers in 1903. But in many years the flowers, which appear before those of all other species of magnolia, are killed by cold winds and frost. So, this species can really only be grown with success in Southern England.

Most of the trees mentioned by Elwes were, in the early 1900's, between 9 and 10.5m (30 and 35ft) tall, but many specimens are now much taller, and are still growing fairly fast. A tree at Wakehurst Place, Sussex, which Mitchell found to be 18.3m × 2.2m (60ft × 7ft 8in) in 1980, had grown to 21.3m × 2.8m (70ft × 9ft 7in) by 1988.

Other trees recently measured show how tall these magnolias have become:-

Borde Hill, Sussex, 26m × 2.1m (85ft × 7ft 3in) 1989
Caerhays, Cornwall, 20m × 2.4m (66ft × 8ft 1in) 1984
Leonardslea, Sussex, 26.8m × 2.1m (88ft × 7ft 1in) 1988
Oak Bank, Gwynedd, 22m × 2.7m (72ft × 8ft 10in) 1984

The subspecies *mollicomata* from the more northern parts of its natural range, has rather deeper pink flowers and produces, when crossed with the type, flowers which are deep purplish pink when young. This cross is called 'Charles Raffill'.

In the USA, the subspecies *mollicomata* seems to be more planted than the type, but it flourishes only in the southern States and in the north west, although it has been known to survive temperatures below -18C (0° F) on eight successive nights in Seattle without injury even to the tips of the branches.

Saucer Magnolia

Magnolia x *soulangiana*

The Saucer Magnolia is a small, hybrid, flowering tree with smooth, light grey bark. There are often several trunks which support a broad, branchy crown. In the spring this is covered with large, showy white to pink flowers, which appear before the oblong, tapering leaves. (See illustration facing p.192)

Origin, cultivation and distribution

The two parents of this hybrid are both species native to China: the Yulan *Magnolia denudata* and the Purple Flowering Magnolia *M. liliflora*.

Both trees occur in the mixed deciduous forests of central and eastern China, mostly on the mid-mountain slopes, and in company with other species such as David's Maple *Acer davidiana*, Golden Larch *Pseudolarix amabilis* and Sweet Gum *Liquidambar formosana*. The Yulan extends further south into the southern part of Hunan Province and is thus the less hardy of the two.

The Yulan, a tree up to 15m (50ft) tall, has been in cultivation in China since around 600 AD and always enjoyed Imperial favour in the old days. It was especially chosen for planting in palace grounds and, despite its southerly origin, seems to have been hardy in Peking with no more than straw ropes wrapped round the bole for winter protection. It was introduced into Britain by Joseph Banks in 1789. Loudon mentions a specimen at Eastwell Park, Kent, which was more than 9m (30ft) tall in 1825. It must have succumbed later because Elwes does not mention it but writes of a tree at Gunnersbury House, Herts., which measured 9.4m × 0.7m (31ft × 2ft 6in) in 1911; and of another at Hornby Grange, Northallerton, Yorks., 9m (30ft) tall, which bore 2,000 flowers in 1896.

The Purple Flowered Magnolia is really a mere bush, never taller than 3.6m (12ft). It was introduced into Britain by Thunberg in 1790 from Japan, where it is not thought to be native. Neither it nor the Yulan are much planted in Britain today because their flowers are so liable to damage by late frosts.

The hybridisation occurred in a Yulan which grew in front of the chateau at Fromont, near Paris, the property of M. Soulange-Bodon, when its flowers were fertilised with pollen from *M. liliflora*. Plants from the resulting seed first flowered in 1826. As the flowers of the Yulan are pure white and those of the other parent white within but vinous purple on the outside, those of the hybrid, *M.* x *soulangiana*, vary greatly in colour from white through shades of suffused pink to deep red, and at 12 to 20cm (5-8in) across are larger than those of either parent; the differences in colour and size being categorised under different clone names.

In Britain the Saucer Magnolia rarely exceeds 3 to 4.5m (10 to 15ft) in height, although Mitchell gives measurements of trees at Riverhill, Hants., as 14m × 1.2m (46ft × 4ft) and 12.8m × 1.2m (42ft × 4ft 2in), and another at Westonbirt, Glos., as 8m × 0.6m (26ft × 2ft 3in); but it is fairly widely planted for its beautiful flowers which appear in April, albeit frequently killed by frost.

In the USA, where it is extremely hardy and can withstand temperatures as low as -32C (-25F), the Saucer Magnolia is a very popular garden tree. Wyman lists no less than 13 varieties, the main differences being in flower colour from pure white to deep purple.

Tulip Tree

Liriodendron tulipifera

The Tulip Tree is the largest and one of the most valuable deciduous trees in eastern North America. It can grow to heights of 45m (150ft) with a beautifully symmetrical bole clad in deeply furrowed bark, often 24m (80ft) or more clear of branches. Its domed crown of up-turned branches bears uniquely saddle-shaped, shiny leaves (glaucous beneath) and, in the spring, greenish-yellow, orange-tinged, tulip-shaped flowers. (See illustration facing p.192)

Distribution and habitat

The Tulip Tree has a wide natural range from Ontario to New York in the north, south to Louisiana and Florida; thus it grows in extremes of climate ranging from severe winters in New England to frost-free winters in Florida. But it grows best where rainfall is distributed over a long growing season and where sites are moderately moist and well drained; very wet and very dry sites are unsuitable. Optimum conditions occur in the lower Ohio Valley and in the southern Appalachians where some of the finest specimens are to be found. In the north these trees rarely occur above 300m (1,000ft), but in the south they may be found up to 1,370m (4,500ft).

The wide range of the Tulip Tree results in its having a great many associated species. On the coastal plains such trees as Sweet Gum *Liquidambar styraciflua*, Swamp Cypress *Taxodium distichum*, Red Maple *Acer rubrum* and Sugar Maple *A. saccharum* are commonly found growing with it; on the lower slopes of the Appalachians, False Acacia *Robinia pseudoacacia*, White Pine *Pinus strobus*, Eastern Hemlock *Tsuga occidentalis* and Black Walnut *Juglans nigra*; and higher up, Black Cherry *Prunus serotina* and Cucumber Tree *Magnolia acuminata*.

Life history

The Tulip Tree starts to produce flowers at 15 to 20 years of age and continues to do so for at least 200 years. Bee pollination is important, and the proportion of viable seed, never very great even in the fairly regular seed years, appears to be directly proportional to the number of bees visiting the flowers each year.

The winged seeds are scattered by the wind. Most seedlings grow quickly, the height after the first year's growth varying between 10 and 30cm (4 and 12in), growth becoming more rapid as the age increases, some 6m (20ft) being attained in 10 years; but records exist of saplings reaching 15m (50ft) in as little as 11 years.This tree coppices readily but such shoots are never as satisfactory as plants grown from seed.

For hardwoods, Tulip Trees attain remarkable sizes. Heights of over 36m (120ft) and girths of up to 1.8m (6ft) are frequently attained in as little as 60 years on good sites, with ultimate heights of 45m (150ft). Elwes records that a tree 58m (190ft) tall was felled in southern Illinois at the turn of the century when it was thought to be the tallest non-coniferous tree ever recorded in the temperate regions of the northern hemisphere. The trunk of another tree felled at about the same time is said to have produced 80,000 shingles.

The Tulip Tree is generally intolerant of shade but because of its fast growth it can overcome competition. It is often a pioneer on cleared sites but is later invaded by oaks and hickories and other more shade-tolerant hardwoods, and is thus usually found growing in mixture.

Timber

The timber of this tree is light yellow to brown in colour, and is soft and easily worked. It is used for many kinds of constructional work, interior finish, boxes, crates, baskets and veneers; and because of the huge sizes available, planks up to 1.5m (5ft) wide can be cut from the logs.

Introduction, performance and use as an exotic

It is not certain when the Tulip Tree was introduced into Britain, but Evelyn thought that Tradescant was the first to bring it over, and one was certainly growing in Bishop Compton's Fulham garden in 1688. A tree still alive at Esher Place, Surrey, bears a label 'This tree was the first to be imported from Virginia in 1685'; according to Gardner, it measured 30m × 7.8m (100ft × 25ft 9in) in 1947. Hunter records that the first Tulip Tree to flower in England was one at Parson's Green, Fulham, which, in 1776, was shaded out by other trees when well past its prime.

Loudon mentions a tree at Killerton, Devon, which was 70 years old in 1843 and was 19m (63ft) tall. Elwes photographed and measured it in 1902, finding it to be 24.4m × 4.5m (80ft × 15ft). It was damaged a year later in a gale and in 1949 was the same height but girthed 4.9m (16ft 10in) when Gardner measured it.

The largest Tulip Tree which Elwes found was at Woolbeding, Sussex, which, in 1903, was 32m × 5m (105ft × 17ft); his photograph shows it as a remarkably handsome, twin-trunked specimen. In 1958 it measured 32m × 6.4m (105ft × 21ft), and, in 1972, Mitchell found it to have grown to 33m × 6.6m (108ft × 21ft 10in), but it was blown down in 1987.

The tallest specimen in 1962 seems to have been at Taplow House, Bucks.; according to Gardner, it was 35m × 5.7m (115ft × 19ft). In 1974 it measured 36.6m × 6m (120ft × 20ft 3in), and may still be the tallest in the country. Other recently measured trees include:-

Glendurgan, Cornwall, 33.8m × 5.9m (111ft × 19ft 8in) 1988
Haffield House, Herefords., 33m × 6m (108ft × 20ft 3in) 1988
Kitlands, Surrey, 33.8m × 5.9m (111ft × 19ft 10in) 1987

Leigh Park, Havant, 32m × 6.4m (105ft × 21ft) 1985
Margam Park, Glam., 31m × 4.6m (102ft × 15ft 6in) 1985
Melbury, Dorset, 22m × 6m (72ft × 20ft 2in) 1989
Stourhead, Wilts., 36m × 6m (118ft × 20ft) 1986

The Tulip Tree seems to do well and be quite long-lived in the south of England on deep, well-drained soils. The tree at Killerton is now over 200 years old and the Esher tree nearly 300. The considerable number of old trees to be found in large gardens and parks and the relatively small number of middle-aged specimens, suggests that although once popular, it declined in favour. But young trees are now quite common and it appears to be regaining some of its lost popularity. A cultivar 'Aureomarginatum', with pale yellow margins to the leaves, is an attractive variation.

As a forest tree it has not been tried, but it might do as well as the native ash on good sites for that species.

In the USA this native tree is frequently planted in parks and large gardens, and the cultivar 'Fastigiatum' is sometimes used as a street tree. It is particularly resistant to insect attack.

Sweet Gum

Liquidambar styraciflua

The Sweet Gum is one of the main components of the deciduous hardwood forests of eastern North America where, in favourable conditions, it can exceed 45m (150ft) in height and 4.5m (15ft) in girth, with a trunk up to 24m (80ft) to the first branch. Mature trees tend to have flat, spreading crowns, but young trees, with their branches springing at acute angles, are usually pyramidal in shape. The star-shaped, five-pointed leaves, which are somewhat like those of the maple, turn wonderful shades of orange and red in the autumn, making these trees an important component of the renowned 'fall' spectacular which the eastern states enjoy. (See illustration facing p.193)

Distribution and habitat

The natural range of the Sweet Gum extends from southern Connecticut in the north to Florida and eastern Texas in the south, ranging from sea level up to 920m (3,000ft) in the Appalachians, and east as far as Oklahoma, Arkansas and south east Mississippi; with scattered locations in Mexico, Guatemala, Salvador, Honduras and Nicaragua. In the Central American areas it is found as high up as 1,520m (5,000ft).

Although tolerant of many soils, it fares best on rich, moist alluvial clays and loams in valley bottoms, and reaches its maximum development in the Lower Mississippi Basin.

In the more northern parts of its range the Sweet Gum grows in mixture with Northern Red Oak *Quercus rubra* and Mockernut Hickory *Carya tomentosa*, and further south with Pin Oak *Q. palustris*, Tulip Tree *Liriodendron tulipifera* (the best soil and site conditions for this tree seem also to be the best for Sweet Gum) and in the extreme south with Nuttall Oak *Q. nuttallii* and Willow Oak *Q. phellos*.

Life history

Seed production in Sweet Gum starts when the tree is 20 to 30 years old and continues for at least 150 years, with good crops each year and bumper ones occasionally. The flowers, which in some areas are regularly damaged by frost, produce spiky fruits which hang, like those of the Plane Tree, singly on long stalks and remain so throughout the winter after shedding their seeds in the autumn.

The seedlings grow rapidly at first, as befits a light-loving species which likes to get above the competition. They are often 60cm (2ft) tall after the first year's growth, reaching 1.3m (4.5ft) in three to five years. In the Mississippi River bottoms these trees frequently top 45m (150ft) with girths of up to 4.8m (16ft). A tree in south Illinois measured 50m (164ft) in height and the 5m (17ft) girth trunk was 24m (80ft) to the first branch. This was at the turn of the century on land now cleared; such trees are now never seen.

The Sweet Gum coppices well and shoots can grow up to 1.5m (5ft) tall in a single season.

There appear to be no records of the Sweet Gum's longevity, but a tree so resistant to disease and insect attack, and reaching the sizes recorded above, must be fairly long-lived, say 200 to 300 years.

Timber

The wood of the Sweet Gum is hard, straight and close-grained and, in colour, bright brown tinged with red, with a thin, white sapwood. It has a satiny lustre and a pleasing, varying figure – a natural beauty excelled by few American woods. It is used, in America, for furniture, flooring, boxes, joinery, veneers and plywood. It was at one time imported into Britain under the name of 'Satin Walnut' (and as 'Hazel Pine' for inferior quality containing an excess of sapwood) and was used for furniture making as well as for barrel plugs. It was also tried as paving blocks for streets, but it rotted too quickly.

Introduction, performance and use as an exotic

The Sweet Gum seems to have been discovered and was certainly first mentioned in 1651 by the Spanish naturalist and physician, Francis Hernandez, who was sent by King Philip II to examine and describe the natural resources of Spanish America. It was introduced to Europe in 1681 by Banister, a missionary collector, who sent a plant home to

Bishop Compton, which was planted in the Palace grounds at Fulham. Loudon, writing in 1838, reported that Sweet Gums had been planted in many collections since the time of introduction and that there were some very fine specimens in the south of England in sheltered situations on good alluvial soils. He mentioned a tree at Syon House, Middlesex, which was then 18m (59ft) tall and 3m (10ft) in girth; and Mitchell records a tree at the same place which, in 1959, was 27.4 × 2.5m (90ft × 8ft 6in) (presumably not the same tree unless Loudon's girth measurements were wrong). In 1976 it measured 29.8m × 2.7m (98ft × 9ft 1in).

Other trees measured include the following:-

Abbotsbury, Dorset, 22.8m × 1.6m (75ft × 5ft 8in) 1980
Jephson Gardens, Leamington Spa, Warwicks., 22m × 1.8m (72ft × 6ft 1in) 1981
Kensington Gardens, London, 22m × 1.8m (72ft × 5ft 11in) 1981
Royal Botanic Gardens, Kew, 28m × 2.1m (92ft × 7ft 3in) 1975
Mote Park, Kent, 25m × 1.6m (82ft × 5ft 6in) 1974
Stratfield Saye, Hants., 28m × 3m (92ft × 10ft 2in) 1986

Even so, the Sweet Gum needs a warmer and moister climate than England provides and generally speaking trees of over 18m (60ft) tall are rare, one reason being its susceptibility to late frost damage which seriously retards growth in the early stages. Nevertheless, it is a fairly common ornamental tree in parks and gardens in the south of England and in parts of the Midlands where it is grown for its neat habit and its handsome foliage. The leaves resemble those of some maples but are set alternately on the stems instead of opposite one another. In England they turn many colours in the autumn from scarlet to deep red, some remaining green, others turning yellow and then purple, but rarely equalling the display which they provide in America.

The gum referred to in the tree's name is a yellowish, fragrant, balsamic liquid which exudes from the bark. Trees grown in the north of its American range and in Britain produce very small quantities, but the gum is produced abundantly from trees in the south of the range and in Central America. This gum resembles and is sometimes substituted for the storax of commerce – a substance collected from another species of Sweet Gum, *Liquidambar orientalis*, native to Asia Minor.

In the USA, this native tree is a very popular ornamental. Although difficult to transplant, it gives little trouble once established; and few other hardwoods match its symmetrical beauty. It is increasingly planted along streets and highways. A cultivar 'Variegata', with leaves coloured irregularly with yellow patches is often planted.

Oriental Plane

Platanus orientalis

The Oriental Plane is a vast, long-lived deciduous tree with a massive, if rather short, bole clad in pinkish-brown bark which flakes away to reveal yellow-green patches. Huge limbs branch off from the trunk and are often so heavy that they rest on the ground half way along their lengths. From their upswept ends and from lesser branches further up the bole a great, spreading crown is formed bearing 15cm (6in) broad indented, five-lobed leaves. (See illustration facing p.193)

Distribution and habitat

The Oriental Plane has been cultivated over so many centuries in the Mediterranean region and in western Asia that its full natural range is difficult to define. It is indigenous in Albania and Greece, growing along the banks of mountain streams in company with Oriental Alder *Alnus orientalis*, and also in similar sites in Crete, Cyprus and Rhodes, as well as in the Lebanon, Syria and Israel. Further east its natural occurrence is in doubt. The great specimens in Persia, Afghanistan and Kashmir, long the wonder of travellers, are almost certainly planted. But planted or wild, they always fare best when growing near water.

Life history

The male and female flowers, borne on the same tree in globules hanging on stalks, appear with the leaves in the spring, the stalks of the females lengthening, after pollination in May, to cover the tree with little spiky balls hanging on long threads. They remain so all through the leafless winter, shedding their seeds in early spring.

Mature Oriental Planes can be enormous and immensely old. Although many of the ancient measurements are supported only by legend, there must be something to them. Pausanias, writing in AD 170, mentions a Plane of extraordinary size and beauty which was said to have been planted by Menelaus, the husband of Helen, and which, when he saw it, was already 1,300 years old. According to Elwes, what was thought to be this very tree, growing in the village of Vostiza, in the Gulf of Lepanto, Greece, measured 42.7m × 11.2m (140ft × 37ft) in 1842, when it was, if Pausanias can be believed, nearly 3,000 years old. When Elwes made enquiries in 1904, it was just a few feet of hollow stump.

Pliny wrote of a Plane in Lycia which had a hollow trunk 24.7m (81ft) in diameter within which the Roman governor dined with 18 guests; and many are the tales of huge, spreading Planes revered by kings and bemedalled by campaigning generals in gratitude for the shade and rest found beneath their branches.

Timber

A tree so revered, planted for the shade it could give, would never be cut for commercial timber. But no doubt small quantities would become available from time to time, when the reddish, brown highly decorative figure on timber cut on the quarter, would be used for ornamental work.

Introduction, performance and use as an exotic

The Oriental Plane was taken to Italy from Greece around 390 BC, and the Romans, who planted it mainly for the great area of shade which it provided, took it to other parts of Europe but not apparently to Britain. However, it did come via Italy, brought by pilgrims in the middle of the 16th century. Loudon lists some specimens, but none of any great size; and it is surprising, in a tree so long-lived, that Elwes does not seem to have found any of the trees mentioned by Loudon.

Elwes tells of a tree at Christchurch College, Oxford, which was reputed to have been planted in 1636. In 1907 its girth beneath a huge, low-down limb was 5.6m (18ft 10in); by 1845 it was 7.2m (23ft 9in). The specimen at Weston Park, Shropshire, in 1875, was said by Elwes to measure 24.4m × 5.5m (80ft × 18ft 6in). A hundred years later Mitchell found its height to have decreased a little but its girth had expanded greatly, for in 1977, it measured 23.4m × 7.3m (77ft × 24ft); and this is the sort of size which many of the old trees in England have now reached. They become more and more massive as their gigantic branches spread out and their crown areas widen. Recent measurements include the following trees:-

Bisberne, Hants., 26m × 6m (85ft × 20ft) 1988
Corsham Court, Wilts., 27.4m × 7.1m (90ft × 23ft 6in) 1989
Royal Botanic Gardens, Kew, 22m × 5.2m (72ft × 17ft 7in) 1987
Melbury, Dorset, 24.4m × 6.3m (80ft × 20ft 9in) 1989
St. Pierre Hotel, Chepstow, 28m × 5.9m (92ft × 19ft 9in); 26.8m × 6.3m (88ft × 20ft 8in) 1988

In southern Europe, where the trees are much older and probably more at home, some immense specimens are to be found, up to 36m (120ft) tall and 9m (30ft) in girth; and some even with multiple trunks and crown spreads of 90m (300ft).

In Kashmir, the 'Chenars' as they are called there, are a great feature of the Mughul Gardens, lining both sides of the terraces and shading the lawns. The grove of 1200 Chenars which Akhbar the Great planted in Nasim Bagh in 1588 is still a feature, with many of the trees still standing with girths of 6m (20ft) and more.

The Oriental Plane was introduced into the USA in colonial times and it has been found to be much less susceptible to twig blight disease than the native *Platanus occidentalis*. It has thus frequently been planted as a shade tree and, being amenable to clipping, it is also used for screens and arbours.

Black Cherry

Prunus serotina

The largest of the American cherries, and the only one of commercial value, the Black Cherry has a spreading crown of ascending main branches from which somewhat tortuous lesser ones extend horizontally. It is held aloft some 18 to 24m (60 to 80ft) on a bole covered in dark purplish-brown scaly bark. After the shiny, oval, pointed leaves open, drooping clusters of white flowers appear, followed by little black pea-sized cherries. (See illustration facing p.193)

Distribution and habitat

The natural range of the Black Cherry stretches from Nova Scotia west through the Canadian Provinces to Lake Superior and Minnesota, south to central Florida and central Texas and on into Arizona, with outposts down the mountain ranges of Mexico and Guatemala. Growing over such an extensive area ensures that this cherry survives a wide range of climatic conditions, from semi-desert to sub-tropical in Mexico and Florida, and the cold, wet climate of Maine and Nova Scotia. But it reaches its maximum development in the southern Appalachians at altitudes between 300 and 760m (1,000 and 2,500ft), where rainfall is regular and ample.

Common associates of the Black Cherry in the north of its range are White Pine *Pinus strobus*, Northern Red Oak *Quercus rubra*, White Ash *Fraxinus americana* and Eastern Hemlock *Tsuga canadensis*; and in the more southern parts, Sweet Gum *Liquidambar styraciflua*, Mockernut Hickory *Carya tomentosa*, Beech *Fagus americana* and Sugar Maple *Acer sachharum*.

Life history

The Black Cherry starts producing seed at about 10 years of age and continues to do so for around 180 years. The flowers come out in March in the south and in June in Canada, where late frosts do great damage. Good seed years occur every three to four years. Most seed falls directly to the ground, although some is dispersed by birds. Germination takes place the following spring when a myriad of seedlings appear beneath the parent trees. Most survive for a year or so, but few cope for longer with the shade and root competition. The more vigorous grow some 10cm (4in) in the first 30 days and, in good conditions, can be 45cm (18in) tall at the end of the first season. Coppice shoots grow very quickly, as do saplings for the first 45 to 50 years. A height of 39.3m (129ft) in a tree 258 years old has been recorded.

This cherry is intolerant of shade and is never abundant except locally where an associate has fallen, leaving a gap. In such conditions the saplings grow quickly and overtop competitors at the edges of the stands. Later they become forked and the crown spreads, allowing other species to overtop the cherries and suppress them.

Timber

The timber is reddish brown, close-grained and, like mahogany, its colour deepens with age. Its beauty, lustre and ease of working make it a valuable wood for cabinet work as well as for veneers and interior trim.

Introduction, performance and use as an exotic

The Black Cherry was probably introduced into Britain in 1724. Loudon gives measurements of a number of trees which had grown to 9 to 10.6m (30 to 35ft) tall in 40 years and were thriving in 1838; as well as one at Chiswick which, after only eight years, had topped 7.6m (25ft). Elwes also saw this tree in 1909 when it had grown to 15.2m × 2m (50ft × 7ft). But the paucity of Loudon's records and, indeed, of Elwes' (only about half-a-dozen in all) suggests that the Black Cherry was never much planted in the early days; nor is it now, despite its handsome glossy leaves, white flower spikes and its shiny black cherries. One reason may be that it is an alternative host of an aphid which carries a virus disease of sugar beet.

Amongst the specimens measured by Mitchell is one at Westonbirt, Glos., which stood at 19.5m × 1.7m (64ft × 5ft 9in) in 1974, one at Tortworth, Glos., which was 16.4m × 1.3m (54ft × 4ft 7in) in 1978, and another at Newick Park, Sussex, which measured 20m × 2.7m (66ft × 9ft 3in) (at 0.9m (3ft)) in 1988.

The specimen at Kew, of the variety *salicifolia*, with narrower, grey-green hanging leaves, measured 16.7m × 2.6m (55ft × 8ft 9in) in 1981.

On the Continent it is planted both as a timber tree and as a nurse for conifers, especially on sandy soils where it is particularly happy, and in some places has become naturalised.

In the USA this cherry, and particularly the graceful variety *pendula*, is rated the best native species for ornamental planting. At one time there were some very large old trees in the eastern States but these have long since gone, for the timber has been highly prized since colonial times, and is still harvested extensively.

Honey Locust

Gleditsia triacanthos

The Honey Locust has a spreading crown, some 21m (70ft) tall, of rather pendulous branches, bearing long, doubly compound leaves, giving the whole a plumed appearance. The branches and some of the twigs, and even parts of the trunk, bristle with bunches of fierce, forked thorns. In spring and summer the male and female flowers appear on separate stalks hanging from the axils of the leaves; the females eventually producing large, twisted pods containing sweet pulp round the seeds, hence the name. The leaves, which open in late spring, turn bright yellow before falling early in the autumn. (See illustration facing p.193)

Distribution and habitat

The natural range of the Honey Locust extends west from central Pennsylvania to south east Dakota and north central Nebraska; south to Texas, east to Alabama and on north east through that state to north Georgia, along the Appalachians to west Virginia, with isolated pockets in north west Florida.

It is typically a tree of the alluvial flood plains of the major rivers and along the banks of streams; but it is also surprisingly resistant to drought and thrives when planted for shelter in the plains. It is very tolerant of both alkaline and saline soils. It is not often found at altitudes higher than 760m (2,500ft), although it can be grown at 1,520m (5,000ft).

Common associates of the Honey Locust are Red Maple *Acer rubrum*, Box Elder *Acer negundo*, Persimmon *Diospyros virginiana*, Kentucky Coffee Tree *Gymnocladus dioicus* and Tupelo *Nyssa sylvatica*.

Life history

The Honey Locust starts to bear seeds at about 10 years of age and continues to do so for 100 years and more, good crops occurring every two or three years. The seeds, which are 2.5-5cm (1-2in) across and borne in pods up to 25cm (10in) long, are thought to germinate most satisfactorily after passing through animals. Growth of both the stem and the tap root of the seedlings is fairly quick – some 45cm (18in) a year for the first seven years.

In the forest this tree will grow to heights of 21 to 24m (70 to 80ft) with girths of 1.8 to 2.8m (6 to 9ft). The largest specimen in America was 42.7m × 5.4m (140ft × 18ft). Maturity seems to be reached at about 120 years; but some seem to be long-lived. Wyman records a tree at Dayton, Ohio, which, when cut down, showed 327 annual rings.

It seems to be intolerant of shade and becomes established only in clearings and then mostly as scattered individuals.

The seed pods, particularly of the southern strains which are grown specially, are a valuable stock fodder.

Timber

The timber is light red in colour and coarse grained, but durable in the soil. It is used for fence posts and railway sleepers.

Introduction, performance and use as an exotic

The Honey Locus was first cultivated in Britain in about 1700 when Bishop Compton planted a specimen at Fulham. It produced fruit 29 years later. But Loudon did not find this tree in 1838, nor did Elwes come across any of the many specimens mentioned by Loudon, most of which were about 18m (60ft) tall around 1838. When recording individual trees in 1910, nearly all of which were about this height, Elwes mentioned a tree growing at Kew which measured 16.4m × 1.9m (54ft × 6ft 10in). This may have been the same tree which Maynard Greville noted as being 19.8m × 1.7m (65ft × 5ft 10in) in 1958. In 1962, the largest Honey Locust was at Wilton House, Salisbury; it measured 20.7m × 1.6m (68ft × 5ft 5in) but this has lost its supremacy, as the latest records show:-

Glasneven Botanic Gardens, Dublin, 18.3m × 2.1m (60ft × 7ft 4in) 1987
Jephson Gardens, Leamington Spa, 19m × 1.5m (62ft × 5ft) 1989
Pampisford, Cambs., 17m × 2.1m (56ft × 7ft) 1988
St. James' Park, London, 21.3m × 1.8m (70ft × 6ft 3in) 1987
The Hall, Avon, 29m × 1.6m (95ft × 5ft 6in) 1989
University Botanic Gardens, Camb., 24.4m × 2.1m (80ft × 7ft 3in) 1984

The Honey Locust is not extensively planted as an ornamental although it has an attractive leaf and flower; this is particularly true of a cultivar called 'Sunburst', which has golden-yellow leaves. Largely because it is not very hardy, it seems to be confined to East Anglia and south east England where it can be found in parks and gardens. It seems to fare better on the Continent where trees up to 24m (80ft) tall are to be found, regularly producing viable seed.

A forest plot of Honey Locusts was planted at Bedgebury, Kent, in 1951, but all the trees were dead within two years.

In the USA, this native tree is widely planted in towns and cities because of its resistance to atmospheric pollution. Wyman lists 12 cultivars including 'Moraine' which is thornless and has been exclusively used as a substitute for elms following the outbreak of Dutch elm disease. It is especially favoured for planting in city streets and in playgrounds where the fiercesome thorns (sometimes as much as 10cm (4in) long) can, in the type tree and other varieties, be a serious hazard.

Pagoda Tree

Sophora japonica

Despite its scientific name, the Pagoda Tree, a lovely flowering species, is native only to China. A straight, cylindrical bole supports the open, rounded crown of tortuous branches, bearing the 15 to 25cm (6 to 10in) long, pinnate leaves each with 9 to 15 leaflets. The trunk, covered in dark brown shallowly fissured bark, often divides into several big branches low down. The white, pea-like flowers cover the crown in late summer. (See illustration facing p.208)

Distribution and habitat

Like so many native Chinese trees, the exact bounds of the Pagoda Tree's natural range are in doubt, so widespread has its planting been; but it is probably wild only in Chihli Province in north China. Like the Tree of Heaven *Ailanthus altissima* and the Foxglove Tree *Paulownia tomentosa*, it appears to be growing naturally in the secondary forests which have regenerated in the Chinese coastal plain. This regeneration follows intense exploitation here as well as on the surrounding foothills, which were also at one time denuded of forest. But in such locations it may well originally have been planted by man.

As a planted tree it is found throughout eastern China from Peking in the north to Hong Kong in the south; and towards the west from Kansu south to Yunnan.

The Pagoda Tree was introduced into Japan so long ago that early travellers to that country deemed it to be indigenous as it was so widespread. Indeed it still is in the temperate and sub-tropical parts.

Life history

The Pagoda Tree does not start to flower until it is at least 30 years old, when the white, pea-like blossoms appear on 25cm (10in) long sprays hanging from the ends of the branchlets. Flowering occurs late in the summer, at the end of July or in August, the petals eventually falling to lie in a white carpet beneath the parent, leaving the seed pods to lengthen to 5cm (2in) and fill with seeds.

Growth is vigorous on good soils, and plants can reach heights of up to 3.6m (12ft) in four to five years. Mature trees are usually between 18 and 24m (60 and 80ft) tall and 2.7 to 3m (9 to 10ft) in girth.

Timber

There appears to be no record of the timber of this tree being exploited commercially in China where it had religious connotations. In the olden days, the Chinese made a yellow dye from the flower buds and a purgative medicine from the bark and roots. In Japan the wood, which is tough and durable, is occasionally used for door frames and such-like items in timber houses.

Introduction, performance and use as an exotic

The Pagoda Tree seems to have been introduced into Britain via France. In 1747, Father d'Incarville sent seedlings from Peking to Bernard de Jussien in Paris. He planted them at Petit Trianon and one of them, according to Bean, was still there and thriving in 1913. In 1753, James Gordon planted some seedlings in his Mile End nursery which he almost certainly obtained from Petit Trianon.

The Pagoda Tree flowered and produced seed for the first time in Europe in 1779, and it was only then that it could be attributed to its correct genus. Subsequently, seeds from France were imported into Britain and from these many of the older specimens are derived.

Loudon mentions a Pagoda Tree at Syon House, Middlesex, which, in 1838, was 17.3m × 0.9m (57ft × 3ft). Elwes also mentions two trees at Syon, each measuring 21.3m × 3.6m (70ft × 12ft), and Mitchell found one of them to be 28m × 4.5m (92ft × 14ft 11in) in 1961; perhaps this was Loudon's tree.

The specimen in the Oxford Botanic Gardens, which was planted about 1818, was 10.6m (35ft) tall when Loudon saw it 24 years later; and when Elwes measured it in 1903 it had grown to 19.8m × 3.6m (65ft × 12ft 3in). By 1962 it was 21.3m × 4.8m (70ft × 16ft) but was felled in 1963. There is still a fine old Pagoda Tree in the Oxford University Parks which, in 1988, measured 17.6m × 3m (58ft × 10ft).

Of the two specimens in the Cambridge Botanic Gardens, which were 15m (50ft) tall in Loudon's time, only one remained when Elwes went there and photographed it in 1904. It then measured 25.3m × 3.3m (83ft × 11ft), but was felled in 1932 when the Gardens in that location were built over. The present-day tree in the Edinburgh Botanic Garden measures 22.8m × 4.8m (75ft × 16ft).

Other more recent measurements include the following trees:-

Anglesey Abbey, Cambs., 18.3m × 2.1m (60ft × 7ft 3in) 1989
Buckingham Palace, 22.8m × 2.1m (75ft × 7ft 2in) 1987
Jephson Gardens, Leamington Spa, 13.7m × 2.7m (45ft × 9ft 4in) 1989
Penrhyn Castle, Gwynedd, 15.2m × 2.2m (50ft × 7ft 7in) 1987
University Botanic Gardens, Camb., 19m × 1.9m (62ft × 6ft 7in) 1989

The Pagoda Tree is a beautiful and interesting ornamental to be found in many of the large gardens in the south of England. The leaves are dark green and glossy with an unusual tint, the branchlets are angular and of a similar colour. The white flowers do not appear until late July or August and thus provide a spectacle long after other trees have become drab. Furthermore, the leaves remain on the tree much longer in the autumn than on many other trees.

On the Continent, where this tree can be found in most countries and in nearly every botanic garden, it flowers more profusely and sets seed regularly.

In the USA, the Pagoda Tree is a very popular ornamental, flowering freely on a wide variety of sites and soils. It is very resistant to air pollution and disease.

It is grown as an ornamental in Australia and Cyprus.

A cultivar 'Pendula' which Fortune saw grafted high up on a tree in Shanghai in 1853, and which has pendulous shoots touching the ground, is sometimes planted, but it never flowers.

Common Laburnum

Laburnum anagyroides

The Common Laburnum is a small tree, rarely more than 9m (30ft) tall, with a bole supporting ascending branches which arch over to form a flat-to-rounded crown of trifoliate leaves. Early in the summer, long pendulous strings of bright yellow flowers appear. The bark is smooth, greenish to pale brown.

Distribution and habitat

The Common Laburnum is a native of Central Europe occurring in upland and mountainous areas, particularly on calcareous soils, up to an altitude of 1,830m (6,000ft). Its range stretches from Lorraine in eastern France through the Jura to Switzerland, Austria and Hungary; the southern limit being northern Italy through Yugoslavia to Bulgaria.

Life history

The yellow, pea-like flowers open in late May and early June, the long festoons of 'golden rain' covering the whole crown. These flowers give way to hairy green pods which ripen a pale brown to shed their black, poisonous seeds during winter. They germinate in April of the following year and growth of the seedlings is rapid.

The Laburnum is not long-lived. Mitchell mentions a tree of 176 years old, which blew down. Nor do these trees grow tall: 9 to 12m (30 to 40ft) being about the maximum height.

Pagoda Tree, Blithfield Hall, Staffs. (p.206)

Tree of Heaven, Hobart, Tasmania (p.212)

Black Locust, Dancer's End, Tring, Herts. (p.210)

Bigleaf Maple, Duncan, British Columbia (p.124)

Norway Maple, Kennel Bank, Dunkeld, Perths. (p.216)

Sycamore, Grindelwald, Switzerland (p.219)

Timber

The wood of Laburnum is very hard and durable, and dark olive brown in colour when dry. It works easily, takes a polish well and makes a first class cabinet-makers' wood, albeit always in short supply.

Introduction, performance and use as an exotic

The Common Laburnum was first cultivated in Britain by Tradescant in 1596, and ever since has been widely planted as an ornamental.

Loudon mentions several trees exceeding 12m (40ft) in height, the largest and oldest in 1838 being at Syon House, Middlesex. He also listed a specimen at Alnwick Castle, Northumberland which, in 1835, girthed 2m (6ft 11in) and contained '46 feet of measurable timber'. Elwes saw this tree in 1905; it was then immense and spreading, and divided into two trunks each 13m (43ft) tall. Below the junction of the trunks the girth at ground level was 3.3m (11ft 2in). In November 1930 it collapsed, but a rooted sprig from the old tree was planted on the same site on 15 November 1933 to mark the coming of age of Earl Percy, later the 9th Duke of Northumberland, and this tree is now over 9m (30ft) tall.

Elwes also describes a four-acre stand of Laburnum at Ickleton Grange, near Cambridge, which had been planted in 1790. The trees had been coppiced at an early age, but the stool growth had formed a thin but pure canopy, beneath which thick grass grew. All natural seedlings, of which there were many, were quickly eaten by rabbits.

Large trees are not common, but Mitchell records one at The Rye, High Wycombe, which in 1980, girthed 1.4m (4ft 9in), one at Kingston Bagpuize, Oxon., which measured 10m × 1.7m (33ft × 5ft 9in) in 1978, and another at Tottenham House, Wilts., measured 11m × 2.1m (36ft × 7ft 4in), with a 1.5m (5ft) bole in 1984.

The Laburnum most commonly planted in Britain and in the USA is a hybrid between *L. anagyroides* and the so-called Scotch Laburnum *L. alpinum*, a tree with much the same distribution as the common species. This hybrid occurred naturally in the Tyrol and was first noticed in 1856. It later appeared in a bed of seedlings at Waterer's Nursery in Surrey, around 1865. The flowers of the hybrid are decidedly more spectacular than those of either parent, combining the large size of the Common with the long racemes of the Scotch. This *Laburnum* x *watereri* is usually called Voss's Laburnum, or Golden Chain in the USA where it is planted in parks and gardens. It is not, though, really hardy in the extreme northern states, and having neither autumn colour nor interesting fruits, as well as only a two-week flowering period, it is of limited appeal there.

Black Locust (False Acacia)

Robinia pseudoacacia

The Black Locust is a deciduous tree with a plume-like crown of bright green, delicately pinnate, compound leaves, studded in the spring with clusters of white pea-like flowers. The twisted thorny branches grow upright from an often forked bowl which has a deeply fissured, reddish brown bark. (See illustration facing p.209)

Distribution and habitat

The natural range of the Black Locust is in two discontinuous parts: in the east in the Appalachian Mountains from central Pennsylvania south to Alabama and Georgia, and including parts of West Virginia, Maryland, Kentucky, Tennessee and the Carolinas, with the western part including south Missouri, Arkansas and east Oklahoma. But it has been planted in almost every state and in many of them is completely naturalised.

In its natural range it grows along streams and on the edges of the forest, as well as in sheltered pockets in the mountains – all areas of humid climate, although when planted on dry sites with high summer temperatures it grows well. Indeed, it is equally accommodating about soil types, limestones being just as acceptable as more acid soils. In the Appalachians it is rarely found growing above 1,070m (3,500ft).

Associated trees on the mountain slopes include Red Oak *Quercus rubra*, Chestnut Oak *Q. prinus*, Tulip Tree *Liriodendron tulipifera* and Pignut Hickory *Carya glabra*; along streams, Black Walnut *Juglans nigra*, White Ash *Fraxinus americana* and Red Maple *Acer rubrum*; and in the southern Appalachians with various pines such as Short-leaved Pine *Pinus echinata* and Virginian Pine *P. virginiana*.

Life history

The Black Locust may start to produce its delicate clusters of white flowers at the early age of six years, but the most prolific age is between 15 and 40 when heavy crops of seed are produced more or less every year. Yet, despite this fecundity, seedlings are rare, largely because of the very hard outer coating of the seed which inhibits germination. Root suckers are by far the most important method of reproduction. Seedling growth is rapid on good sites; up to 1.2m (4ft) a year has been recorded, although half that is more usual. Root suckers have an exceptional rate of growth 2.4 to 3m (8 to 10ft) a year being not unusual. This tree is short-lived, rarely exceeding its century; but heights of 23m (75ft) are quite often reached in 50 years and trees 30m (100ft) tall and up to 2.7m (9ft) in girth are quite common on good sites.

Although very intolerant of shade, the rate of growth of the Black Locust is such that it is often able to compete with its more intolerant companions.

Timber

The timber is greenish yellow to dark brown in colour, heavy, hard and tough, as well as being remarkably durable in the ground. It is used for stakes, posts, poles, tool handles; and at one time it was supreme in America for insulators for telephone and transmission poles.

Introduction, performance and use as an exotic

The Black Locust was one of the first American trees to be introduced into Europe by means of seeds sent to Jean Robin, gardener to King Henry IV of France, in 1601. The first known tree was one planted by Robin's son, Vespasian, arborist to King Louis XII, in the Jardins des Plantes in 1635. Until very recently the stump of this tree was still producing sprouts.

The date of introduction into Britain is not really known, but may well have been earlier than into France, as Parkinson, writing in 1640, mentioned a specimen in Tradescant's Lambeth garden which had grown to 'an exceeding height'.

A specimen at Kew, planted in 1762, girthed, according to Gardner 4m (13ft 6in) in 1950, and is still there, a hulk 4.6m (15ft 6in) in circumference.

William Cobbett returned from America in 1819 with seed of the Black Locust, and much enthusiasm for the tree's performance. He began selling plants from 1832 onwards, accompanied by his own brand of propaganda extolling the tree's virtues as a timber producer. In the end he sold over a million trees. A few of these are still alive but well past their prime. As a forest tree it never 'took on', probably because of its suckering habits, its thorns, its brittleness in winds and the difficulty of working its timber.

The finest tree which Elwes knew of and of which he took a photograph, was at Frogmore, Isle of Wight. It measured 26.8m × 4.3m (88ft × 14ft 7in) in 1908, an girthed 4.8m (16ft 2in) when Gardner saw it in 1948. The specimen at Bowood, Surrey, which Elwes found to be 27.4m × 2.6m (90ft × 8ft 8in) in 1908, was 30m × 2.7m (99ft × 9ft 2in) when measured by Maynard Greville in 1957, and the tallest in Britain at that time.

More recent measurements include the following trees:-

Burr Castle, Co. Offlay, 21.3m × 3.3m (70ft × 11ft 3in) 1985
Butleigh Court, Som., 25m × 3m (82ft × 10ft 1in) 1984
Claremont House, Surrey, 28m × 5.1m (92ft × 17ft 3in) 1986
Hampton Park, Surrey, 26m × 6.7m (85ft × 22ft 3in) 1984
Longleat, Wilts., 21.3m × 3.6m (70ft × 12ft 4in) 1986

Old Rectory, Sudborough, Northants., 17m × 5.3m (56ft × 17ft 8in); 29m × 2.5m (95ft × 8ft 5in) 1987
Royal Botanic Gardens, Kew, 15.2m × 4.9m (50ft × 16ft 9in) 1989

The Black Locust is very common in Britain, especially on sandy soils, but it finds no favour as a forest tree. Its chief merit was in being the provider of durable fencing stakes, for which it was planted on some estates, especially in Norfolk. However, this has been eclipsed by advances in preservative treatment of other more easily workable timbers.

As an ornamental tree it has merit and it is often planted for its general good shape, delicate leaves and handsome flowers, although its suckering can be tiresome. More popular, perhaps, is the cultivar 'Freesia' which has beautiful golden leaves.

On the Continent, where the Black Locust grows very vigorously, it is in many places almost naturalised. In France it is often seen as a severely pruned street tree.

In the USA, this native tree is frequently planted as an ornamental, but two serious pests, the locust borer and the locust leaf miner, often disfigure it. Apart from the type tree, several cultivars are popular: 'Decaisneana' with pink flowers, 'Semperflorens' which flowers throughout the summer, and 'Umbraculifera' with a dense rounded head.

Tree of Heaven

Ailanthus altissima

The Tree of Heaven is a large deciduous species. A native only of China, it has huge compound leaves, up to 90cm (3ft) long with 7 to 20 leaflets borne on generally ascending but sinuous branches to give a flattened, domed crown. This is supported by a smooth, grey bole, streaked by paler-coloured, longitudinal fissures. (See illustration facing p.208)

Distribution and habitat

The natural range of the Tree of Heaven is not easy to define, so great has been the deforestation in northern China, but it is probably wild only in the mountains of Chihili Province. Nevertheless, it is now also a component of some of the secondary forest areas which have grown up following the destruction of the temperate deciduous broadleaved forest of the north China coastal plain and around Peking. Here seedlings and young trees can be found growing naturally on the once denuded foothills in company with many other species including Pink Siris *Albizzia julibrissin*, Juniper *Juniperus chinensis*, Walnut *Juglans regia*, Foxglove

Tree *Paulownia tomentosa* and Pagoda Tree *Sophora japonica*. Throughout most of China the Tree of Heaven is common in cultivation.

Life history

The male and female flowers are borne on separate trees, the females producing an abundance of winged seeds which hang in bright red bunches amongst the green leaves. These give a pleasing display before being dispersed by the wind. The tree also reproduces itself by root suckers, large collections of them arising around a single bole.

Growth is rapid and heights of up to 30m (100ft) are reached in 50 or 60 years, with girths of up to 4m (13ft).

Timber

Despite the large sizes available, the timber, being soft and very prone to decay in the open, is only used locally for the construction of small houses and for firewood. It is said to work well and take a polish, and has been used in a very small way for furniture in Europe and America.

Introduction, performance and use as an exotic

The Tree of Heaven was introduced into Britain by seeds sent to the Royal Society from Nanking in 1751 by Father d'Incarville, a Jesuit missionary. These were sown by Miller in the Chelsea Botanic Gardens and by Philip Webb at Busbridge, near Godalming, Surrey. When Loudon visited this latter place in 1834, the original tree planted by Webb had been felled but several suckers were growing from its roots.

It seems that this tree is not long-lived in Britain. The tallest specimen recorded by Loudon, at Syon House, Middlesex, which was 21m (70ft) tall in 1838, had grown to 30m (100ft) by 1880, and was dead when Henry looked for it in 1906. The two trees of equal size – 23.7m × 2.7m (78ft × 9ft) – at Steventon, Berks., measured by Henry, and regularly producing seed at the turn of the century, could not be found when Gardner looked for them in 1947. And another specimen at Fulham, photographed by Henry, and measuring 24.4m × 2.7m (80ft × 9ft) in 1902, had to be felled in 1932 following gale damage. The tree at Barton, Suffolk, planted in 1826, measured 16.7m × 1.5m (55ft × 5ft) in 1904. Despite surviving the winter of 1860, producing flowers in 1861 and viable seed in 1868, it had gone by 1913. The tallest specimen which Henry found was at Belton Park; it measured 25.3m × 1.8m (83ft × 6ft) in 1904, but again Gardner could not find it when he visited the place in 1949.

More recent measurements include the following:-

Abbey Gardens, Bury St. Edmunds, 24.4m × 3.3m (80ft × 11ft 3in) 1984
Bishops Palace, Wells, 20m × 3.2m (66ft × 10ft 9in) 1986

Canford School, Dorset, 21.3m × 3.8m (70ft × 12ft 10in) 1989
Hampstead Heath, London, 29m × 2.1m (95ft × 7ft 3in) 1989
Highnam Court, Glos., 24.4m × 3.1m (80ft × 10ft 6in) 1988
University Botanic Gardens, Camb., 21.3m × 3.1m (70ft × 10ft 5in) 1989
University Park, Oxford, 25m × 2.8m (82ft × 9ft 6in) 1988
Westonbirt, Glos., 29.6m × 1.9m (98ft × 6ft 6in) 1989

The Tree of Heaven is exceptionally resistant to air pollution and is thus a prized tree for street planting in the south of England. For this purpose the female tree should be used, as the male flowers have an unpleasant smell.

Although the climate of Britain restricts this tree to the south of England, it is extensively planted on the Continent; and in southern Europe it has in places become naturalised, especially in Italy where it is well able to withstand drought.

As a forest tree in Europe it has failed, largely because it is very intolerant of shade and cannot compete with native species.

In Australia it has been found to be a valuable tree for checking erosion because of the profuse production of suckers. It is also planted there as a street tree and as an ornamental and shade tree in Cyprus, Israel and South Africa.

The Tree of Heaven was introduced into the USA in 1794 and soon became highly popular for street planting because of its ability to thrive on almost any soil and in air laden with soot, gas and smoke. It can also withstand occasional inundation by salt water near coasts subject to hurricanes; and it is not susceptible to insect pests.

In many parts of the country it has become naturalised.

A variety *erythrocarpa* with bright red fruits, and a cultivar 'Pendulifolia' with very large, drooping leaves are also planted.

Bigleaf Maple

Acer macrophyllum

This tree is the largest of the American maples and the only one of commercial value growing in the west of the country. It is a component of the great forests of the north west and can often be seen as a broad-crowned, spreading tree on the edges of the Douglas Fir stands. The boles of open-grown trees tend to fork into stout ascending branches forming a broad and rounded crown at a height of about 18m (60ft); but in close-grown stands, where the trees are drawn up by competition, heights of up to 45m (150ft) have been recorded. The bark is thin but deeply furrowed in old trees, although smooth grey-brown at first in saplings. The five-lobed leaves are very large for a maple – 15 to 30cm (6-12in) across – deep green and shiny above, paler beneath, turning orange in the autumn, but not giving such a spectacular display as many other maples, especially those in eastern North America. (See illustration facing p.209)

Distribution and habitat

The natural range of the Bigleaf Maple extends over 20 degrees of latitude from the southern tip of Alaska in the north to southern California in the south. It is mainly a tree of the lower-lying coastal areas but ascends to 300m (1,000ft) in the north of the range in British Columbia and to as much as 2,140m (7,000ft) in the coastal mountains of southern California.

It grows on a variety of soils but fares best on the deep alluvial types, especially near streams, and reaches its maximum development in Oregon and Washington, particularly around Puget Sound where there are occasional limited pure stands with boles festooned with mosses and ferns, so moist and humid are the sites which it likes best.

In the more northern part of its range the Big-leaf Maple is found growing with Douglas Fir *Pseudotsuga menziesii*, Western Red Cedar *Thuya plicata*, Grand Fir *Abies grandis*, Western Hemlock *Tsuga heterophylla*, Vine Maple *Acer circinatum* and Red Alder *Alnus rubra*. Further south such trees as Californian Laurel *Umbellularia californica*, Madrone *Arbutus menziesii*, Redwood *Sequoia sempervirens*, Californian Live Oak *Quercus kelloggii* and White Alder *Alnus rhombifolia* are found with it.

Life history

Bigleaf Maples start to produce flowers at about 10 years of age and continue to do so for 150 years and more; open-grown trees often bearing huge crops of seed year after year and providing adequate natural regeneration for silvicultural operations. The Bigleaf Maple stands up very well against its competitors: it has more copious seed production, seedling growth is more rapid, and in early life it is fairly shade tolerant – and remains so even up to 60 years on good sites. For these reasons, it can form nearly pure stands although eventually it finds itself overtopped and pushed to the forest edge by Douglas Fir and other conifers.

The keys (seeds) which are produced in pairs set at angles of 45 to 90 degrees, are extremely difficult to store at any temperature.

At the extremities of its range the Bigleaf Maple is often short stemmed, gnarled and barely 9m (30ft) tall, but on optimum sites, in Oregon it can be a huge tree with a crown spread of 18m (60ft) and more, and some 6m (20ft) in girth. When heavily shaded in dense stands it can top 30m (100ft) with about half its trunk free of branches. Cut stools coppice freely and the shoots grow rapidly.

Timber

The timber of this maple is light brown with a reddish tint, firm and fine-grained. It is used in north west America for furniture, flooring, boats and boxes.

Introduction, performance and use as an exotic

The Bigleaf Maple was discovered by Archibald Menzies in the late 1700s. Although plants were probably sent to Britain as early as 1812, the first seeds from which the main introduction was made were sent home to the Horticultural Society by David Douglas who, at the time, forecast that 'it will at some future time constitute one of the most ornamental forest trees in England'. For some reason this prophecy has not come about, for the tree has not really found favour and is not commonly planted in Europe. Although it is hardy, with only a slight tendency to be cut back by early frosts, grows vigorously and has large and handsome leaves, it has never become popular; perhaps because its autumn colouring is not as intense as other maples. Even so it deserves more attention, although as a forest tree it cannot compare with Sycamore. Some large trees are to be found in Britain, as Mitchell's measurements show. When young the tree grows quickly, just as in America. Miles Hadfield had a specimen in his Herefordshire garden which, after planting out from a pot in 1962, was 2.7m (9ft 4in) tall two years later.

Recent measurements include the following trees:-

Brodick Castle, Arran, 19.8m × 2.3m (65ft × 7ft 9in) 1988
Trebartha, Cornwall, 28m × 2.3m (92ft × 7ft 8in) 1981
Trinity College, Dublin, 15.8m × 3.3m (52ft × 11ft 3in) 1985
Westonbirt, Glos., 26.8m × 3.3m (88ft × 11ft 1in) 1988

The Bigleaf Maple is a popular roadside and shade tree on the Pacific coast of north west America, but it does not seem to grow well in the eastern states. Wyman mentions a narrow, upright specimen in the University of Washington Arboretum at Seattle which has been named 'Seattle Sentinel', but it is not yet being propagated commercially; nor for that matter have any other cultivars.

Norway Maple

Acer platanoides

The Norway Maple is a large, deciduous tree, often 24m (80ft) tall, generally with a short bole covered in a pale grey, rough bark shallow fissured longitudinally. The trunk supports a wide, domed crown of typically maple-shaped, bright green leaves 12cm (5in) long, separated into five sharply pointed lobes, thin and shiny smooth on both sides. They turn bright yellow in the autumn. (See illustration facing p.209)

Distribution and habitat

The natural distribution of the Norway Maple stretches from the Pyrenees in the west right through Central Europe to the Caucasus and northern Iran in the east. It reaches southern Norway, Sweden and Finland in the north and on into western Russia, and to northern Italy and the Balkans in the south. It is not found growing naturally in northern France, Belgium, Holland and Britain.

This maple is never found in pure stands of any size but is generally solitary or in small groups with other species such as Sycamore *Acer pseudoplatanus* and Beech *Fagus sylvatica* in Central Europe and with Birch *Betula pendula*, Spruce *Picea abies* and Aspen *Populus tremula* in northern areas. In Switzerland it ascends up to around 920m (3,000ft) but in the Caucasus and further east it is found as high up as 1,830m (6,000ft). In northern latitudes it grows mainly at low elevations and more often than not near coasts.

Life history

The bright yellow flowers appear in profusion massed in erect bunches on the bare twigs in April and May before the leaves open. These bunches become pendulous as the 'keys', the winged seeds set in pairs and widely divergent, develop and ripen in September and October.

Seedling growth is rapid and saplings can be as much as 3.5 to 4.5m (12 to 15ft) tall in 3 years; later, growth slows up. Mature trees can be as much as 27m (90ft) tall, with girths of 3m (10ft) or so. But they are not long-lived as trees go, although a planted specimen in Scotland, according to Mitchell, is known to be at least 250 years old.

Timber

The timber, yellowish-white to pink in colour, is very hard with no distinct heartwood and is used for turnery, furniture, brush backs and kitchen utensils.

Introduction, performance and use as an exotic

According to Loudon, the Norway Maple was introduced into Britain in 1683. He mentions one at Kew which stood at 23m (76ft) in 1838 after being planted 70 years. This was the tallest he knew of in England. He also lists a number of specimens up and down the country of varying sizes from 15 to 21m (50 to 70ft) tall, all of which seem to have grown about a foot a year over their lives, except one in Ireland which at 60 years was 23.7m (78ft) tall.

The tallest known to Elwes was at Cassiobury Park, Herts. It measured

29m × 4.1m (95ft × 13ft 9in) in 1906, but Gardner reported it dead by 1952. It was not so shapely as two others which he photographed at Westgate, near Emsworth, Hants., and at Park Place, Henley-on-Thames. The former, which had a crown spread of 102 paces in circumference and girthed 4.2m (14ft) in 1906, had increased its girth to 4.9m (16ft 6in) by 1954. The latter measured 24.4m × 2.9m (80ft × 9ft 9in) in 1905. All these trees must now have gone, for the tallest which Maynard Greville could find was one at Pampisford, Cambs. which measured 24.7m × 2.7m (81ft × 9ft 2in) in 1952 and 26m × 3.1m (85ft × 10ft 7in) in 1988.

Other trees recently measured include:-

Batsford Park, Glos., 29.8m × 3m (98ft × 10ft 3in) 1987
Castle Fraser, Aberdeen, 26m × 3.6m (85ft × 12ft 3in) 1983
Doune House, Inverness, 24.4m × 3.6m (80ft × 12ft) 1989
Eastnor Castle, Herefords., 22.8m × 3.5m (75ft × 11ft 9in) 1989
Hafodonas, Clwyd, 26m × 3.3m (85ft × 11ft 3in) 1984
Westonbirt, Glos. (Waste), 32m × 3.3m (105ft × 11ft) 1988

The Norway Maple is very common in Britain, seeding itself readily on sandy soils. It is planted in gardens and parks, often in avenues along streets and in groups on roundabouts. It is particularly pleasing in the autumn when the leaves turn a bright yellow.

In America, too, this maple has long been planted, especially in the eastern states, and in the Pacific region. It has proved to be one of the best exotic ornamentals, being very hardy and less prone to disease and insect attack than the native maples. For town planting it also has an advantage over the native species in that it does not ultimately grow so large and it fares well on poor soils.

As many as 20 cultivars have been produced from this maple of which the most generally planted, both in Europe and America, is 'Crimson King'. It arose in Orleans in 1946 and has leaves which are deep ruby red, with a 'wet look'. It, or something very like it, is also sold under the name of 'Goldsworth Purple'. Another is 'Schwedleri', the young leaves of which are pink, turning dark green and later orange red. Also often planted in the States as a street tree is 'Erectum' with a narrow, fastigiate habit; and 'Globulosum', a low-growing, rounded tree, best grafted high up on the type root stock.

Sycamore

Acer pseudoplatanus

The Sycamore, the largest of the maples, is, when open-grown, a huge, spreading deciduous tree with a dense, domed crown of large, five-lobed leaves which are dark green with a paler underside. The massive trunk, often swollen at the base, has a pinkish brown bark, scaling off in irregular strips which curl up at the edges. (See illustration facing p.209)

Distribution and habitat

So extensively has the Sycamore been planted in Europe over many centuries that its exact limits of natural distribution are not easy to define; but it is a tree of the central mountain chain – the Pyrenees, the Alps and the Carpathians – and of the hilly districts which radiate from it. The western limit is the Cantabrian Mountains of Spain, and it extends eastwards almost as far as the Caspian Sea, north as far as the Hartz Mountains and south into the mountains of Thessaly and Epirus in Greece.

Pure natural stands are rare and of small extent. More often it is found as isolated trees or in small groups and as a minor component in the great Beech *Fagus sylvatica* and Silver Fir *Abies alba* forests of the European mountains. However, it only competes successfully with these species on rich soils in valleys and ravines. It is found at altitudes between 300 and 1,220m (1,000 and 4,000ft), in places reaching up to 1,620m (5,300ft), dwindling to a sub-alpine shrub near the tree line.

Life history

The yellowish flowers, borne in mid-April on long, hanging racemes, develop into winged seeds or keys set in pairs. These are scattered far and wide by winter gales to germinate and grow in immense profusion. Growth at first is very rapid – 45cm (18in) a year for the first four years – giving them a fair start over many competitors. Growth then slows down but continues up to a great age.

Although top heights may not often exceed 30m (100ft), old trees develop gigantic girths with enormous spread of crowns. Elwes shows a photograph of a sycamore growing at 1,340m (4,400ft) in Melchtal in the Unterwalden Canton of Switzerland, with a girth of 8.8m (29ft) (12m (40ft) at the swollen base), and with a branch 2.7m (9ft) in girth 3.6m (12ft) from the ground. As to age, there was a tree at Truns in the Bernese Oberland under which the Grey League (one of the three bodies which eventually formed the Canton of Grisons) were sworn in in 1424. Its last remnants disintegrated in a storm of 1870.

Timber

The fine textured timber is white to yellowish, with a natural lustre. It is moderately hard and easy to work and is used in the turnery trade for making brush handles and domestic utensils. It has a high resistance to wear as a flooring timber. Some trees produce a timber with a wavy grain which is greatly sought after for veneers.

Introduction, performance and use as an exotic

Just as its natural distribution is difficult to define with accuracy, so the date of the Sycamore's introduction into Britain is far from certain. Clement Reid, in his *Origin of British Flora*, suggested, without any evidence, that it was brought here by the Romans; but the first mention of the tree in any literature is in Turner's *Herbal* of 1551. Gerard, in 1597, mentions it as being a stranger to England and only found in the walks and places of pleasure of noblemen. Some authorities suggest that it may have been brought to Scotland in the first instance due to the historic intimacy of that country with France. But whenever it came, it has been remarkably successful in establishing itself throughout the entire realm, and there are some enormous specimens of great age to be seen, many of them in the north of England and Scotland.

None of those listed by Loudon was recorded by Elwes, but the tree at Cobham Hall, Kent, which, in 1905, he found to be 32m × 5.3m (105ft × 17ft 9in), was measured by Maynard Greville in 1953 when it was 35.6m × 5.7m (117ft × 19ft 2in) and the tallest in Britain at that time. Another specimen at Drumlanrig Castle, Dumfries, which Elwes photographed and measured in 1904 as 32m × 5.9m (105ft × 19ft 6in), had grown to 33.8m × 7m (111ft × 23ft 3in) when measured by Mitchell in 1989. Elwes also mentioned a tree at Birnam, Dunkeld, which Hunter supposed was 1,000 years old, and which girthed 5.8m (19ft 8in) in 1883. Mitchell measured it in 1974 when it was 30.8m × 7m (101ft × 23ft 4in), and again in 1989 when it was no taller but girthed 8m (26ft). Other recent measurements include the following trees:-

Biel, E. Lothian, 35.6m × 4.2m (117ft × 14ft 4in) 1985
Lennox Cove, Midloth., 40.2m × 3.6m (132ft × 11ft 10in) 1985
Martyrs' Tree, Tolpuddle, girth 5.7m (19ft) (1986)
Newbattle Abbey, Midloth., 26m × 5.4m (85ft × 18ft) 1989
Plas Newydd, Anglesey, 19m × 6.8m (62ft × 22ft 6in) 1987
Tyninghame, E. Loth., 32m × 7.3m (105ft × 24ft) (at 0.9m (3ft)) 1989

The Sycamore is very hardy and resistant to strong winds, particularly salt-laden gales off the sea, which considering its continental origin, is surprising. It is much planted in shelterbelts and for shelter round hill farms. In northern Britain it is an important hedgerow tree.

As a forest tree its ready regeneration from natural seedlings makes it very acceptable as a minor component in woodlands, especially on private estates, although damage from bark-stripping by grey squirrels restricts its use in some areas.

As a hedgerow tree in northern Britain it often exceeds in size and useful life-span native species such as oak, ash and wych elm.

For ornamental planting, perhaps the most handsome is the cultivar 'Brilliantissimum' which is much slower growing than the type but produces a dense, round crown of bright pink leaves which turn orange in mid-summer and green later on. Perhaps the most romantic of the cultivars is 'Corstorphine' in which the leaves emerge bright yellow and remain so until July. The original tree is a survivor of an avenue which led up to an old manor house belonging, since 1376, to the Forrester family in what is now the Corstorphine area of Edinburgh. The 2nd Lord Forrester, who fell foul of Cromwell and then quarrelled with his sister-in-law, is said to have murdered her at the foot of this tree in 1679. According to Elwes it measured 18.6m × 3.3m (61ft × 11ft)in 1905, but Mitchell found it to be 16.7m × 3.7m (55ft × 12ft 5in) in 1975, so it is probably going back in the crown.

A cultivar 'Purpureum', which occurred in Jersey in 1828 has purple undersides to its leaves; and 'Variegata' has pale green leaves with yellow splotches.

The Sycamore or Great Maple was introduced into the USA in colonial times. Although it cannot begin to compete with native maples for autumn colour, its resistance to salt-laden winds makes it a popular subject in seaside gardens, and the cultivar 'Erythrocarpa' is well-liked because of its red fruits. A yellow cultivar 'Worleei' which retains its colour longer than 'Corstorphine' is also planted.

Smooth Japanese Maple

Acer palmatum

This Japanese maple is a small deciduous tree, up to 15m (50ft) tall with a regular, rounded crown of slender, brownish-green branches, somewhat upswept and sinuous, all supported by a smooth greyish white bole. The leaves are 7cm (3in) wide and lobed with five to seven points. (See illustration facing p.224)

Distribution and habitat

This maple is found growing naturally in China, Japan and Korea. In China it is a minor component of the mixed forests in the mountains of northern Chekiang at between 1,220 and 1,520m (4,000 and 5,000ft), where it is associated with such trees as Chinese Dogwood *Cornus kousa*, Golden Larch *Pseudolarix amabilis* and Japanese Cedar *Cryptomeria jap-*

onica. In Taiwan it accompanies other deciduous trees near the upper limit (around 1,830m (6,000ft)) of the evergreen broadleaved forest of that island. In Japan it is distributed throughout the mountainous parts of the country, and it is found in its multiplicity of varieties and cultivars in almost every Japanese garden.

Life history

The pinkish-red female flowers, in bunches on erect stalks, open in April and May at the same time as the leaves, producing the typical maple 'keys', set in pairs at obtuse angles, and hanging on drooping stalks until they fall in October.

On good sites early growth of the seedlings is rapid, some reaching 90cm (3ft) in height in the first season.

Mature trees are rarely more than 13.5 to 15m (45 to 50ft) tall, with girths of about 90cm (3ft).

Timber

In Japan the hard, heavy, white, close-grained timber is used for furniture, musical instruments, wood carving and internal decoration.

Introduction, performance and use as an exotic

The Smooth Japanese Maple was discovered by Thunberg in 1783 and was originally introduced into Britain from Japan in 1820, but the main introduction was by Wilson from central China many years later. Even so, it is not the species tree which is now planted in almost every garden in the temperate world but one of the 250 cultivars which have been developed, mainly in Japan.

More than 300 years ago the Japanese started to collect the unusual and beautiful forms of this maple from the natural stands – seedlings, sports and variants. By 1710, 36 varieties had been named and it was fashionable to collect and cultivate as many as possible in the hope that others would appear both naturally and by hybridisation. By 1882 cultivars and varieties numbered 250, but many of these disappeared during the second world war. However, renewed interest during the last 30 years has resulted in the reappearance of many choice varieties and the emergence of many new ones, the list again numbering about 250.

This legion of separate forms, nearly all differing chiefly in either leaf shape or colour, as well as in size, is extensively grown throughout Europe and North America on almost every kind of soil and situation where exposure is not too severe.

Large species trees are rare in cultivation. Recent measurements show some of the best:-

Megginch Castle, Perths., 12.2m × 1.9m (40ft × 6ft 6in) 1986
Minterne, Dorset, 12.8m × 1m (42ft × 3ft 4in) 1988
Muncaster Castle, Cumbria, 17m × 1.2m (56ft × 4ft 6in) 1984

Trebah, Cornwall, 17m × 1.2m (56ft × 4ft 6in) 1985
Wakehurst Place, Sussex, (Lawn), 12.2m × 1.2m (40ft × 4ft 4in) 1988
Westonbirt, Glos., 14m × 1.6m (46ft × 5ft 6in) 1988

Of the many cultivars, some of the most popular include 'Atropurpureum', dark purple turning deep red in the autumn; 'Osakazuki' turning a brilliant scarlet; 'Senkaki', the Coral Bark Maple with crimson shoots, which show up beautifully in the winter, and small slightly yellowish leaves; as well as several varieties with delicately cut leaves of which *A. p. dissectum*, with a weeping habit, is a very handsome example.

Red Maple

Acer rubrum

The Red Maple has a wide range in the east of North America. The dark grey bark, flaking on the surface into plate-like scales, covers a considerable bole in old trees, supporting a broad oval crown, often as tall as 30m (100ft), of upswept, whip-like branches. In the spring the whole crown becomes ruby red as the clusters of flowers come out before the bright green, five-lobed leaves which become interspersed later with clusters of red keys. In the autumn the leaves turn a brilliant scarlet, tinged here and there with gold. (See illustration facing p.224)

Distribution and habitat

The Red Maple ranges throughout the eastern United States from the Atlantic Ocean to the Prairies, also taking in the maritime provinces of Quebec and Ontario in Canada, and south to Florida and to Texas. Thus, it grows in a wide variety of climatic conditions, but reaches its maximum development in well drained, moist sites at low elevations, such as the lower Ohio and Mississippi valleys. It is common in mountain country, even on dry ridges in the south.

With such a wide range, the Red Maple is associated with up to 70 different commercial tree species. In the north with such trees as Paperbark Birch *Betula papyrifera*, Sugar Maple *Acer saccharum*, Black Cherry *Prunus serotina* and Red Spruce *Picea rubens*: and in the south with Slash Pine *Pinus elliottii*, American Elm *Ulmus americana*, Green Ash *Fraxinus pennsylvanica*; as well as, in swampy areas, with Swamp Cypress *Taxodium distichum* and Water Tupelo *Nyssa aquatica*.

Life history

The Red Maple is one of the first trees to flower in the spring, some trees being entirely male, others female and some bearing flowers of both sexes, although usually on separate branches. The seeds are dispersed by the wind in June and July, germinating as soon as they fall.

The seedlings quickly grow a 30cm (1ft) tap root and a 30cm (1ft) tall shoot in the first season and then for the next few years they grow about 60cm (2ft) a year, maintaining a very high proportion of the species in mixed stands.

Under good conditions this maple may grow to 36m (120ft) tall, with a girth of 4.5m (15ft); but it is not long-lived, 150 years being an average span. It is usually a pioneer species and, being longer-lived and more tolerant of shade than many of its competitors, it often becomes an intermediate species in many stands, and later a major understorey species, as in Pennsylvania, under climax stands of Red Pine.

Timber

The light brown timber is hard and close-grained, and easily worked. It is used for cabinet work, flooring, veneers, gunstocks and turnery.

Introduction, performance and use as an exotic

According to Loudon, Tradescant was cultivating the Red Maple in his Fulham nursery as long ago as 1656. Bishop Compton planted one in his Fulham garden towards the end of the 17th century, a tree which Miller found to be 12m (40ft) tall in 1793, but which had died by the time Loudon visited the place in 1809. He listed a number of specimens most of which, by 1838, seem to have been 12m (40ft) tall at ages of about 45 years.

Elwes found none of them, but those which he did list in the early 1900's were much taller than Loudon's. One at Bagshot, among a number of fine specimens of this species which he photographed, was 25m × 2.8m (82ft × 9ft 6in) in 1902. Its girth had increased to 3m (10ft 4in) when Gardner measured it in 1954 and may be the same tree which Mitchell found to girth 3.6m (12ft 4in) in 1982. At Whitton, near Hounslow, there was a tree which Elwes also photographed and measured at 24.4m × 2.5m (80ft × 8ft 5in). He computed it to be 150 years old in 1904, but as Loudon did not mention it, this age may be exaggerated.

More recent measurements include the following trees:-

Eastnor Castle, Herefords., 22m × 2.4m (72ft × 8ft) 1989
Melbury, Dorset, 18.3m × 2.5m (60ft × 8ft 7in) 1989
Trewidden, Cornwall, 17m × 2.7m (56ft × 9ft 4in) 1979
Woodhill, Perths., 21.3m × 2.5m (70ft × 8ft 4in) 1986

The Red Maple is certainly very hardy in Britain and, although it does not colour in this country as brilliantly as in America, its generally neat habit, attractive leaf shape, and the early appearance of its wine-red to yellow autumn colouring make it a very desirable ornamental.

In the USA, this native maple is popular as a lawn tree because of the

Smooth Japanese Maple var. 'Osakazuki', Exbury, Hants. (p.221)

Red Maple, Vermont (p.223)

Silver Maple, Dancer's End, Tring, Herts. (p.225)

Horse Chestnut, Station Road, Tring, Herts (p.227)

Yellow Buckeye, Lichfield, Staffs. (p.230)

Indian Horse Chestnut, Simla. Langurs eating fruit. (p.232)

red flowers which appear in the spring and its early autumn colouring, despite its being rather brittle and liable to damage from snow and ice storms. The densely upright cultivar 'Columnare' is much planted and, when brilliant red in the 'fall', is a remarkable sight.

Silver Maple

Acer sacharinum

Open grown, the Silver Maple develops a short trunk dividing into a number of large, ascending branches to form a broad, rounded crown 23m (75ft) tall. The five-lobed, deeply indented leaves are light green above and silvery white beneath and flash in the sunlight as they are agitated by the wind. In dense forest the trunks are branch-free up to 9 to 12m (30 to 40ft) and reach heights of 36m (120ft). The bark of old trees is thin and grey, broken into broad scales which flake off. In young trees it is smooth and slightly tinged with red. (See illustration facing p.224)

Distribution and habitat

The Silver Maple is a tree of lowland deciduous forests in east and central North America, its natural range extending west to Minnesota and south to north west Florida and Louisiana, as well as to the southern parts of New Brunswick and Ontario in Canada. It is rare near the Atlantic coast and absent from the higher Alleghany and Appalachian Mountains. Although found on most soils, it is more numerous where there is a good moisture supply throughout the year, and thus it tends to grow best on the banks of slow-flowing rivers, on the borders of swamps and on other flat and gently sloping land. In the prairie States it thrives only along river courses. It reaches its maximum development in the Ohio River basin and is more common there than anywhere else.

This maple is rarely found in pure stands but is usually scattered throughout the forest where suitable conditions exist and where its principal associates are Red Maple *Acer rubrum*, Basswood *Tilia americana*, Swamp White Oak *Quercus bicolor*, Cottonwood *Populus deltoides*, Plane *Platanus occidentalis*, Green Ash *Fraxinus pennsylvanica* and Black Walnut *Juglans nigra*.

Life history

Forest grown Silver Maples start to flower at about 35 years of age. The flowers appear, before the leaves open, as yellow bundles on the bare twigs, eventually producing the typical maple 'keys', but hairier than most species. Because of the seeds' extreme sensitivity to drying out they must germinate immediately or not at all. Growth is rapid, the seedlings reaching 30 to 90cm (1 to 3ft) in the first summer. Stumps up to a foot in diameter will produce very vigorous coppice shoots.

Growth continues to be rapid for up to 50 years. Open grown trees of 23m (75ft) in height and forest grown specimens of up to 36m (120ft) are common, with girths as much as 3.6m (12ft). But it is not a long-lived tree, some 125 to 150 years being an average life span.

Although usually found in mixed stands, the Silver Maple is fairly shade-tolerant and is sometimes a major component of the forest. Occasionally it may invade open, swampy areas as a pioneer species, but eventually reverts to its normal status as a sub-climax species.

Timber

The timber is hard, brittle and close-grained. It is used for cheap furniture, flooring and interior finish.

Introduction, performance and use as an exotic

This maple was introduced into Britain by Sir Charles Wager in 1725, but has never become as popular as it deserves, although in Europe, particularly in Germany and Belgium it is widely planted as an ornamental and roadside tree.

Its rapid growth and the beauty of its deeply lobed leaves, which, having silvery undersides, flash and sparkle in the slightest breeze, make it a very attractive tree. Furthermore, it seems to be tolerant of many different sites and soil types.

Elwes recorded a specimen in Kent which was 23.4m × 2.7m (77ft × 9ft), as well as many others between 15 and 18m (50 and 60ft) tall, whilst Mitchell notes that the tallest in Britain in 1963 was a tree at Leighton Knolls, Shropshire, which he measured as 26m × 3m (85ft × 10ft). Since then he has found others considerably taller, as his recent measurements show:-

Henrietta Park, Bath, 22m × 3.3m (72ft × 11ft) 1986
Newby Hall, Yorks., 20m × 3.3m (66ft × 11ft 10in) 1989
Royal Botanic Gardens, Kew, (Jap Gate), 29m × 4m (95ft × 13ft 3in) 1989
Westonbirt, Glos., 31m × 2.7m (102ft × 9ft 4in) 1988
Wisley, Surrey, 29m × 2.7m (95ft × 9ft 4in) 1988

There is a number of cultivars of the Silver Maple with a variety of leaf shapes and sizes of which 'Laciniatum' is becoming increasingly popular.

This native maple is widely planted as a street and roadside tree in the USA, but it is generally more brittle than the closely related *Acer rubrum* and is thus not so popular.

Horse Chestnut

Aesculus hippocastanum

The Horse Chestnut is one of the largest flowering trees of the temperate world. Often over 30m (100ft) tall with a huge, symmetrical, domed crown of large, palmately compound leaves, the leaflets spreading like the fingers of a fat hand. The branches are upswept at first and then dip down, only to turn up again at the tips. In winter, the fat sticky buds are very distinct. The stout bole has reddish-brown bark which peels in long plates that become detached at both ends, giving it a shaggy appearance. The candle-like white flowers cover the crown in the spring. (See illustration facing p.225)

Distribution and habitat

The natural range of the Horse Chestnut is limited to parts of Albania and Greece where it is found in moist valleys in the mountains at altitudes of between 920 and 1,220m (3,000 and 4,000ft) in company with such trees as Grecian Fir *Abies cephalonica* var. *appollis*, Walnut *Juglans regia*, Hop Hornbeam *Ostrya carpinifolia*, Oriental Plane *Platanus orientalis* and Grey Alder *Alnus incana*.

Life history

The white flowers on upright pyramidal panicles 25cm (10in) tall, appear in masses all over the crown like candles in May and June. They give way to large, globular, green and spiky fruits which split open in September to reveal one to three large, shiny, mahogany-brown seeds or 'conkers'.

Growth is rapid in the early stages and young trees can add 60 to 90cm (2 to 3ft) to their height annually, eventually reaching 30m (100ft) or more. Some specimens of over 300 years old are known, but many start to deteriorate at half that age.

Timber

The timber, which is close-grained and white in colour, is soft and lacking strength and durability. Even if it were available in large quantities it would have few uses; it does not even burn well.

Introduction, performance and use as an exotic

In Loudon's time the Horse Chestnut was still believed to be a native of Asia, and he says that it 'passed from Asia to Europe about the middle of the 16th century'. In a sense this is true because the tree was first raised in Europe from seeds sent to Clusius in Vienna by the Imperial Ambassador in Constantinople in 1576. It was introduced into France in 1615 by Bachelier, again by Constantinople seed; and it seems that a tree was growing in Tradescant's Lambeth garden in 1633, so it probably came to Britain about the same time as it went to France. But it was not until 1879 that its natural distribution in Albania and Greece, and not in Asia, was satisfactorily proved by Heldreich. He found it growing in remote, totally uninhabited areas which established to his satisfaction that the tree was really wild. Later opinions corroborated this.

Since the early days after introduction, the Horse Chestnut has been extensively planted in Britain. Loudon, writing in 1838, lists many specimens between 80 and 100 years old. The tree at Syon House, Middlesex, which, in 1838, was 24m (80ft) tall and some 3m (10ft) in girth was found by Elwes in 1905 to be 28.3m × 4.5m (93ft × 15ft). The tallest specimen which he knew of was at Petworth House, Sussex. It was 35 to 36.5m (115 to 120ft) tall in 1906, and Mitchell found it to be 38m × 5.7m (125ft × 19ft 2in) in 1961 and the tallest in the country at that time. More recent measurements include the following trees:-

Castle Ashby, Northants., 30m × 5.3m (100ft × 17ft 9in) 1988
Hurstbourne Priory, Hants., 36m × 6.7m (118ft × 22ft 2in) 1984
Kingston Lacey, Dorset, 35m × 5.4m (115ft × 18ft 2in) 1983
Preston Hall, M'sex, 26m × 5.7m (85ft × 19ft) 1987
Spetchley Park, Worcs., 21.3m × 5.4m (70ft × 17ft 10in) 1985
Trevorich, Cornwall, 29.8m × 4.5m (98ft × 14ft 10in) 1989

That such a magnificent flowering tree, very hardy and quick-growing (most of them grow 30cm (1ft) a year for at least 60 years) should be popular is not surprising. Indeed, were they not so common, they would be a tourist attraction. Whether standing singly in a field or park, the great symmetrical domed crown amass with white candles, or in avenues where they merge together in continuous lines of undulating green, they are one of the great tree spectaculars of the temperate zones.

Where conker-hunting by children is a hazard, such as in parks, the fruitless cultivar 'Baumannii' is often planted. It originated in Geneva about 1820.

An attempt to improve this spectacular tree by finding one with red flowers (gilding the lily, one might say) was made in France around 1818 when a fertile hybrid between *Aesculus hippocastanum* and the American Red Buckeye *A. pavia* grew into a reasonably sized tree with red flowers, *Aesculus* x *carnea*. It has been extensively planted, but it is a poor tree compared with the Horse Chestnut, with dull, very dark green leaves, an indifferent crown and highly susceptible to canker. However, a cultivar of this hybrid 'Briottii', also produced in France, in 1958, has brighter red flowers and glossier leaves.

The Horse Chestnut, introduced into the USA about 1828, was once widely planted there, but its susceptibility to disease, its brittle branches and its lack of autumn colour have resulted in its replacement by the red hybrid, both the type and 'Briottii', as well as by the Damask Horse

Chestnut, the cultivar 'Plantierensis' which has pink flowers with red or yellow throats. Both the Horse Chestnut and the red hybrids have a distinct advantage over the native Buckeyes as far as floral display is concerned.

Japanese Horse Chestnut

Aesculus turbinata

This exclusively Japanese tree is very similar in general habit to and is obviously a very close relation of the European Horse Chestnut. It has thick, scaly bark on a sturdy bole and a domed crown of heavy branches with substantial twigs ending in sticky buds during the dormant period. But although the leaves are the same palmate shape, their size is remarkable and distinctive; they are as much as 60cm (2ft) across, shiny green and turn a bright orange in the autumn to provide a magnificent display.

Distribution and habitat

The natural range of the Japanese Horse Chestnut stretches from South Hokkaido in the north, where the tree is rather scattered, south right through Honshu, where it is very common in the northern half, and down to Shikoku and Kyushu, becoming less common the further south it is found. It is mainly confined to the mountains at elevations up to 450m (1,500ft) in Hokkaido and up to 1,675m (5,500ft) in the south of Honshu, growing in mixture with Japanese Cedar *Cryptomeria japonica*, Keaki *Zelkova serrata* and oaks such as *Quercus crispula*.

It has, for centuries, been planted as an ornamental in Japan's towns and cities.

Life history

The white flowers, smaller but otherwise very similar to those of the European species, borne on upright stalks, open in May and June producing, by October, brown warty fruits, not spiky like the European relative, but containing the same lustrous brown seeds. In their native forests these trees attain top heights of between 24 and 30m (80 and 100ft), with girths as much as 4.2m (14ft) and trunks clear of branches up to 12 or 15m (40 or 50ft).

Timber

In a country such as Japan where there are so many fine hardwood trees, the timber of the Horse Chestnut is not greatly valued because it is soft and rather weak. Nevertheless, it sometimes has a wavy figure which, combined with its subtle reddish colour when seasoned, give it a place in internal decoration. Elwes saw a plank of this wood in Osaka measuring 4.5m (15ft) long and 1.5m (5ft) wide, with a wavy grain.

Introduction, performance and use as an exotic

The date of introduction of the Japanese Horse Chestnut into Britain is uncertain, but was probably around 1880. Elwes had a tree at Colesbourne which flowered for the first time in 1901, but he does not record having seen any notable specimens.

Bean, writing in 1914, mentioned two trees at Westonbirt, one of which flowered regularly and produced viable seeds, but he does not give any measurements. Mitchell noted four trees there (possibly including the same two) in 1988. They measured the following: by the house; with broken top, 15.2m × 2.4m (50ft × 8ft 4in); in the arboretum, 22m × 2.5m (72ft × 8ft 8in); Willes, 24.4m × 1.6m (80ft × 5ft 9in); at the entrance, 25m × 2.1m (82ft × 6ft 11in).

This tree does not appear to grow so quickly as the European species, nor does it flower so profusely; but the size of its leaves and the colour they assume in the autumn make it an attractive and interesting subject for a large garden or park.

Wyman gives this tree no mention, so it cannot be a commonly planted species in the USA.

Yellow or Sweet Buckeye

Aesculus octandra

The Yellow Buckeye has a domed crown of small, rather pendulous and twisty branches bearing 15cm (6in) long palmate leaves with five slender-stalked leaflets. In the spring it produces yellow flowers in erect candle-like clusters. Open grown, this tree reaches about 18m (60ft) in height, but under forest conditions it can be up to 30m (100ft) tall with a long, clean trunk. The bark of old trees is dark brown, divided by shallow fissures and separating on the surface into small, thin scales.

In the autumn this tree joins in the general display of colour with its leaves turning a yellowish orange. (See illustration facing p.225)

Distribution and habitat

The natural range of the Yellow Buckeye extends westwards from north Pennsylvania along both sides of the Ohio River to south east Illinois and southwards along the Alleghany and Appalachian Mountains to east Tennessee and north west Georgia. It grows mainly in valley bottoms and along the banks of streams in the northern part of the range but, further south, on the high mountain slopes where it is commonly found at elevations of between 450 and 1,820m (1,500 and 6,000ft), often in considerable concentrations but mainly forming only 15 to 30 per cent of the canopy in company with other trees such as Sugar Maple *Acer sachharum*, Basswood *Tilia americana*, Beech *Fagus americana* and Yellow Birch *Betula alleghaniensis*.

Life history

The fruits produced from the yellow flowers of the Yellow Buckeye are round, smooth, leathery capsules containing, usually, two large seeds about 3.5cm (1.5in) across. These ripen in September and are dispersed by gravity, animals and water. They lie in the ground all winter and germinate in the spring, developing a large taproot very early on.

This tree is shade-tolerant and, being one of the first among its associates to leaf out in the spring, it quickly becomes established, although over very limited areas. This tendency towards relatively restricted distribution is probably due to the difficulty of disseminating such large seeds and, perhaps, an inability of the seedlings to establish themselves except on the best sites. On such sites the Yellow Buckeye can grow to a considerable size. The largest reported was in the Smoky Mountains; it was 26m (85ft) tall, with a crown spread of 16.5m (54ft) and a girth of 4.5m (15ft). Another in Ohio also girthed 4.5m (15ft).

Timber

The timber of the Yellow Buckeye is light, soft, close-grained and creamy-white in colour. It is difficult to split, and is used for artificial limbs, joinery, kitchen ware and pulp; and occasionally as saw-timber for limited uses.

Introduction, performance and use as an exotic

The Yellow Buckeye was probably introduced into Britain somewhere around 1764, but it has never grown so well or been so popular as the European Horse Chestnut *Aesculus hippocastanum*. Loudon, writing some 72 years after the first introduction, lists a dozen or more measured specimens in England and Scotland of which the largest was one at Syon House, Middlesex, which was 12m (40ft) tall. Elwes found this same tree to be 19.8m × 1.2m (65ft × 4ft 4in) in 1904, and Bean recorded it as 'nearly 70 feet (21m)' in 1933. But such a large tree is exceptional, although recent measurements include some trees taller than this in several parts of the country:-

Bath Botanic Gardens, 22m × 3.3m (72ft × 11ft 1in) (at 60cm (2ft)) 1984

Easton Grey, Glos., 22m × 1.1m (72ft × 3ft 10in) 1988
Kinfauns Castle, Perth, 21.3m × 1.9m (70ft × 6ft 8in) 1985
Melbury, Dorset, 21.3m × 1.9m (70ft × 6ft 6in) 1989
Much Hadham Rectory, 26m × 2.6m (85ft × 8ft 9in), 22m × 2.1m (72ft × 7ft 4in) 1989
St. Margarets, Twickenham, 21.3m × 3m (70ft × 9ft 10in) 1989
Stowe Park, Bucks., 18.3m × 2.4m (60ft × 7ft 10in) 1986

Despite the handsome rounded crown and yellow flowers this tree is only infrequently seen in the larger gardens and parks in Britain, and these are often grafted specimens, growing on the rootstock of *Ae. hippocastanum*. This is mainly done to perpetuate the considerable number of varieties which have sprung up with different shades of colour in their flowers, varying from pink to scarlet with reddish veins offsetting the yellow.

Indian Horse Chestnut

Aesculus indica

The Indian Horse Chestnut is a tree of the north west Himalayas and one of the largest and most handsome flowering trees of the temperate regions. In old trees the massive bole is clad in bark which peels off in large flakes that remain attached at the top, hanging in tatters. The domed crown is made up of smooth, dark green, palmate leaves, the leaflets radiating from the centre like fingers of a hand. The yellow to pinkish flowers are set 30 to 40 at a time on 37cm (15in) upright stalks, and the mass effect of hundreds of them, candle-like all over the crown, is remarkable. (See illustration facing p.225)

Distribution and habitat

The natural range of the Indian Horse Chestnut is in the lower mountains of the north west Himalayas stretching from the Indus in the north west to the western border of Nepal in the south east, at elevations between 1,220 and 2,440m (4,000 and 8,000ft).

The tree occurs in mixture with other species in the moist, temperate, deciduous forests of this region, growing in damp hollows and depressions, along hill streams and on the gentler slopes.

Associated species include Walnut *Juglans regia*, Bird Cherry *Prunus padus*, Pindrow Fir *Abies pindrow*, Deodar *Cedrus deodara* (the latter two at higher elevations), as well as a number of evergreen oaks *Quercus* spp. and elms *Ulmus* spp.

Life history

The flowers appear in April in India (much later in Europe) and produce smooth, leathery, brown fruits containing one to three dark brown, shiny seeds about 2.5cm (1in) in diameter. These are dispersed by gravity, birds and monkeys, germinating soon after falling to the ground; any drying out seems to be fatal.

Growth is rapid when there is sufficient light and the trees eventually reach considerable sizes. Open grown they have short, massive trunks sometimes girthing 12m (40ft) and reach top heights of 24m (80ft) and more. When drawn up in the forest they can be as much as 45m (150ft) tall.

The Indian Horse Chestnut is never an important component of the forest, but tends to be well scattered amongst the large number of other species which make up the rich forests of this region.

Timber

The white, rather soft timber is used locally for building, water troughs, turnery and tea boxes.

Introduction, performance and use as an exotic

The Indian Horse Chestnut was introduced into Britain in 1851 by Colonel Henry Bunbury who brought seeds to his home at Barton, Suffolk, and raised plants from them. By 1904, when Elwes measured two of these, they had grown to 20m × 2.3m (66ft × 7ft 9in) and 19.8m × 2m (65ft × 7ft 2in). The smaller one, which had divided into two trunks, flowered for the first time when seven years old. Ten years after planting, having survived the terrible winter of 1860, one of the trees produced fruit from which four plants were raised. Elwes saw and photographed the parent tree in full flower on 24 June 1905.

There were few other large specimens in Britain at the turn of the century but gradually this tree has started to gain the popularity it deserves. Among recent records are the following trees:-

Hidcote, Glos., 18.3m × 2.6m (60ft × 8ft 9in) (bole 1.8m (6ft)) 1983
Lindridge Park, Devon, 17m × 2.4m (56ft × 8ft 5in) (bole 1.8m (6ft)) 1985
Royal Botanic Gardens, Kew, 17m × 2.2m (56ft × 7ft 7in) 1989
Townhill Park, Hants., 21.3m × 2.6m (70ft × 8ft 9in) 1985
Westonbirt, Glos., 21.3m × 2.1m (70ft × 7ft 4in) 1988

Given a moist soil and a reasonably sheltered situation, the Indian tree can rival, if not surpass, the European species in the beauty and profusion of its flowers, although perhaps not in size. As the European tree flowers in April and the Indian in July, the advantage of planting both species is obvious.

Even more floriferous is the cultivar, 'Sydney Pearce', with green as opposed to red leaf stalks, and a sturdier habit, which competes on even better terms with the European species. It occurred at Kew in 1928. The original tree near the main entrance measures 12.8m × 1.5m (42ft × 5ft).

Golden Rain Tree

Koelreuteria paniculata

The Golden Rain Tree or Pride of India is a medium-sized tree. The trunk has a rough, scaly, purplish-brown bark and supports a domed crown of somewhat sinuous branches. The dark green, pinnate leaves are 30cm (1ft) long and, in late summer, huge panicles of bright yellow flowers appear on the branches. (See illustration facing p.240)

Distribution and habitat

This tree is a native of north east China, Korea and, probably, Japan. In China it is found in the hills bordering the Great Plain of the north coast in company with such species as the Tree of Heaven *Ailanthus altissima*, Pagoda Tree *Sophora japonica*, and Foxglove Tree *Paulownia tomentosa*. Further south, in the provinces of Shensi and Kansu, it grows in the oak forests on the lower slopes of the mountains below 1,220m (4,000ft) in association with Walnut *Juglans regia*, Chinese Juniper *Juniperus chinensis* and Chinese Thuya *Thuya orientalis*. In Japan, it is widely distributed but some authorities think it may not be indigenous and was probably introduced from China in ancient times. In China, too, it is extensively planted and in former days it was placed around the graves of feudal princes.

Life history

The bright yellow flowers, 1.2cm (0.5in) across and in clusters of three, are borne on huge spreading panicles over 30cm (1ft) long. They appear in August and are followed by papery, bladder-like fruits containing black, pea-sized seeds which germinate readily. Seedling growth is fairly rapid, and the average annual increase in height over the first 20 years is about 30cm (1ft). Mature trees rarely exceed 18.3m (60ft) in height.

Timber

The timber is little used, but the seeds are strung together to make necklaces and the leaves and flowers are used for making black and yellow dyes.

Introduction, performance and use as an exotic

The Golden Rain Tree was first introduced into Europe when D'Incarville sent seeds from China to Paris in 1747. It was first cultivated in Britain by the Earl of Coventry at Croome, Worcestershire, in 1763.

Loudon mentions a number of specimens in southern England of which the tallest was at Ham House, London, measuring 13m (42ft) in 1838. Neither this tree nor any of the others he listed were found by Elwes who records that the finest specimen he saw was one at Waterer's Nursery, Woking, which measured 12.2m × 1.8m (40ft × 6ft) in 1911. He noted another at Windsor Castle of the same size which he records was covered with flowers in July 1912. But Gardner could not find this tree in 1948. So, it seems that it may not be a long-lived tree in Britain, and it is certainly not hardy in the north. Most specimens nowadays are to be found in southern England or East Anglia where they are very attractive subjects with their large leaves, pink when unfolding, dark green and feathery throughout the summer and turning crimson in the autumn. Their masses of yellow blossoms appear later than those of most flowering trees and are succeeded by the attractive pink fruits.

Recent measurements include the following trees:-

Battle Abbey, Sussex, 15.8m × 1.6m (52ft × 5ft 6in) 1986
Chelsea Physic Gardens, 11m × 2.4m (36ft × 8ft) 1989
Oxford Botanic Gardens, 15.8m × 1.2m (52ft × 4ft 4in) 1989
Sandford Park, Cheltenham, 19m × 1.5m (62ft × 5ft) 1989

In the USA, where the Golden Rain Tree was introduced at about the same time as into Britain, it is widely planted in the less cold regions. With Laburnum, it is the only truly yellow blossomed species which will grow in the Arnold Arboretum. Its quick growth, ready resistance to drought and insect attack, and its tolerance of many soil types, as well as its floral beauty, make it a popular ornamental. It seeds freely there and in many places it has become naturalised. A cultivar 'Fastigiata' is also used as a street tree, particularly in the Ohio Valley.

Silver Lime

Tilia tomentosa

The Silver Lime is a large, deciduous tree, up to 30m (100ft) tall. It has sharply ascending branches arising from the bole, to form a broadly domed crown. The bark is pale grey with smooth, flat ridges. The heart-shaped leaves are dark green and crinkly above and covered with dense, pale grey hairs (tomentum) beneath. The white undersides of the leaves flash all over the crown when agitated by the wind. (See illustration facing p.240)

Distribution and habitat

The natural range of this tree extends from south east Europe into Asia Minor. From its northern limit in north Hungary it ranges south through Yugoslavia to a southern limit of Laconia in Greece; and eastwards through Romania and Bulgaria to the Crimea.

It is usually a component of mixed woodland in company with such trees as Hungarian Oak *Quercus frainetto* and other deciduous species. Occasionally it may be found in small, pure stands.

Life history

The pale yellow, fragrant flowers appear late, in July, seven to ten hanging on a pendulous stalk and eventually giving way to little nut-like, woody fruits containing two or three seeds.

Early growth is rapid and is maintained to give height increases averaging up to 30cm (1ft) a year for the first 50 years. Mature trees vary between 23 and 30m (75 and 100ft) in height and about 3.6m (12ft) or so in girth.

Timber

The light, fine-grained timber is used, in its native lands, for turnery and household ware and for items for which a timber without taste or smell is important, such as apiary equipment.

Introduction, performance and use as an exotic

The Silver Lime was first grown in Britain at Gordon's Nursery, Mile End, London, in 1767. Loudon recorded some good specimens at Kew in 1838, and one at Walton-on-Thames which was 18m (60ft) tall. Another at Croome, Worcs., only 30 years old, was over 15m (50ft) and girthed 1.8m (6ft). Nearly all the trees he mentions were growing on relatively shallow chalky soils.

Elwes rated the finest to be the one at Albury Park, Surrey, which he photographed and measured as 30m × 4m (100ft × 13ft 6in) in 1906. Another at Harpsden Rectory, Oxon, which Henry measured as 24.4m × 1.9m (80ft × 6ft 8in) in 1907, had increased in girth to 2.8m (9ft 6in) when Gardner measured it in 1949. Recent measurements include the following trees:-

Buckland House, Brecon, 28m × 5.6m (92ft × 18ft 9in) 1988
The Rye, High Wycombe, 29m × 1.9m (95ft × 6ft 6in) 1980
Tortworth Church, Glos., 35m × 4.5m (115ft × 15ft) 1986
Westonbirt, Glos., 29m × 3.4m (95ft × 11ft 5in) 1988

The Silver Lime is common in towns and cities in Britain as well as in large gardens and in estate policies up and down the country. Its symmetrical shape, large size and spectacular display of flashing white leaves give it a special attraction.

In the USA, the Silver Lime, which was introduced in colonial times, is often planted as a park, street or highway tree, even though the hairy undersides of the leaves are prone to collect soot and grime. Its ability to withstand both heat and drought are added advantages.

Black Gum or Black Tupelo

Nyssa sylvatica

The Black Gum is a tree of eastern North America notable for its neatly shaped crown of simple, shiny, dark green leaves which turn a wonderful scarlet in the autumn. The trunk, with its dark grey bark deeply and longitudinally fissured, extends right up into the top of the tree. Perpendicular, spiky shoots grow from the many short and slightly upturned main branches to form the dense crown. (See illustration facing p.240)

Distribution and habitat

The natural range of this tree is bounded in the north by a line from south west Maine, south Ontario and central Michigan; by the Atlantic coast down to Florida in the east, and across to Texas; and in the west down through Illinois, central Mississippi to south east Oklahoma.

It is adapted to a wide variety of sites from the damp river-valley bottoms of the southern coastal plain, to the dry upper slopes of the Appalachians at altitudes of 920m (3,000ft), as well as the dry uplands of the eastern Great Plains. But it is most at home and attains its greatest sizes on the loams and clays of the lower slopes of the Appalachians in South Carolina and Tennessee.

The Black Gum is never dominant in the forest but is found in mixture with various species in the several parts of its range. In the north it associates with, amongst others, the Black Ash *Fraxinus nigra*, Elm *Ulmus americana* and Red Maple *Acer rubrum*; in the more central areas with Sweet Gum *Liquidambar styraciflua*, Scarlet Oak *Quercus coccinea*, Tulip Tree *Liriodendron tulipifera* and Eastern Hemlock *Tsuga canadensis*; and in the south with Slash Pine *Pinus elliottii*, Southern Magnolia *Magnolia grandiflora* and Beech *Fagus grandifolia*.

Life history

The inconspicuous greenish flowers appear in the spring when the leaves are still opening, and the female ones develop into little, plum-like, blue-black fruits, 1.3cm (0.5in) long, which fall in the autumn, the enclosed single seeds being dispersed by small animals. They overwinter in the damp soil and germinate in the following spring.

On good sites the top height of mature trees is often about 36m (120ft) with girths of up to 3.6m (12ft).

At low elevations, the Black Gum is usually found growing in mixture with other trees and is most often in the intermediate crown class; but in the mountains it frequently has a place amongst the dominants with which it is associated.

Timber

The light brown timber is very durable but has a twisted grain and is difficult to split. It is used for flooring, gunstocks, pistol grips, turnery and boat keels.

Introduction, performance and use as an exotic

The Black Gum seems to have been introduced into Britain before 1750. Loudon mentions the date 1739 and records a tree at Richmond which was 13.7m × 0.4m (45ft × 1ft 4in) in 1838. He also mentions a specimen at Stratfield Saye, Hants. as being 9m (30ft) tall in 1838, and Elwes photographed and measured this same tree as 22.5m × 1.5m (74ft × 5ft) in 1897. It first produced seed in 1906, and when Gardner measured it in 1947 it was 21.3m × 2m (70ft × 6ft 9in). In 1978 Mitchell found it to have grown to 24.4m × 2.1m (80ft × 7ft 5in). Other trees recently measured include:-

Bicton, Devon, 17m × 1.9m (56ft × 6ft 7in) 1983
Cannizaro Park, Wimbledon, 20m × 1.5m (66ft × 5ft) 1988
Chatsworth, Derbys., 19m × 1.8m (62ft × 6ft 4in) 1982
Royal Botanic Gardens, Kew, 19m × 1.8m (62ft × 6ft 3in) 1987
Sheffield Park, Sussex, (planted, 1909), 12.2m × 1.5m (40ft × 5ft 3in); 12.8m × 1.5m (42ft × 5ft 1in); 12.2m × 1.5m (40ft × 5ft) 1988

As an ornamental tree the Black Gum's chief merit is its brilliant autumn colouring, but the tree is not common in Britain being seen mainly in large gardens and collections in southern England, and rarely further north than the Midlands.

Because of a very straggling root system it is difficult to transplant with success.

This native tree is a popular ornamental in the USA, especially in the eastern states, although, as elsewhere, it is difficult to transplant. Its pyramidal habit and its spectacular autumn colouring, as well as its relatively small size, make it a useful garden subject.

Cider Gum

Eucalyptus gunnii

The Cider Gum is a tree of high altitudes and is thus usually somewhat stunted. It bears its domed crown of thick, firm but narrow, pointed, evergreen leaves on upswept, sinuous branches. The short trunk is covered in bark which is rough at the base, but smooth higher up, showing grey beneath pinkish, peeling strips. (See illustration facing p.241)

Distribution and habitat

The Cider Gum occurs only in upland areas of Tasmania at elevations between 920 and 1,220m (3,000 and 4,000ft), most common on steep, broken country and exposed ridge tops where soils are often poor and shallow. Associated trees include such interesting conifers as King Billy Pine *Athrotaxis selaginoides* and Celerytop Pine *Phyllocladus aspleniifolius*, as well as Myrtle Beech *Nothofagus cunninghamii* and a number of eucalypts such as Tasmanian Snow Gum *E. coccifera* and Urn Gum *E. urnigera*.

Life history

The fluffy white flowers develop into pear-shaped, glaucous capsules, with flat, unrimmed tops, containing a huge number of tiny seeds. Growth of seedlings is, like most eucalypts, rapid: 30cm (1ft) in the first year and 0.9 to 1.8m (3 to 6ft) in the second, although on some of the harsher sites growth is very much slower. In Tasmania on the best sites top heights of 24m (80ft) are attained, but 9 to 18m (30 to 60ft) trees are more commonly seen.

There do not appear to be any records of life span, but as in most alpine trees, small gnarled specimens can linger on for centuries.

Timber

The timber is not exploited commercially because of the inaccessibility of most stands.

Introduction, performance and use as an exotic

The Cider Gum, so called because it yields an agreeably sweet sap, was discovered by Sir Joseph Hooker in central Tasmania in 1840. It was introduced into Britain in 1846, the first tree being planted at Kew. It was also the first Australian tree to be planted out of doors in Europe, but it did not seem to prosper, for by 1863 it was only 6m (20ft) tall after being cut down to the ground by frost in several severe winters.

None of the trees mentioned by Elwes was more than 15m (50ft) tall, probably because the older ones had suffered frost damage. However, Mitchell records a specimen at Trebah, Cornwall, where severe frosts must be rare, which measured 33.5m × 3.4m (110ft × 11ft 8in) in 1959; and another at Sidbury Manor, Devon, which was 22.8m × 4.1m (75ft × 13ft 9in) in 1961. Other recent measurements include the following trees:-

Castle Kennedy, Wigtons., 28m × 3.3m (92ft × 11ft 10in) 1989
Culzean Castle, Ayr, 19.8m × 3.6m (65ft × 12ft) 1989
Inverewe, Ross & Cromarty, 36m × 2.4m (118ft × 8ft 2in) 1986
Menabilly, Cornwall, 25m × 3.1m (82ft × 10ft 6in) 1989
Wakehurst Place, Sussex, 22m × 2.9m (72ft × 9ft 9in) 1988

In 1852, the Marquis of Salisbury brought back some supposedly *E. gunnii* seed from Tasmania from which hundreds of plants were raised by the Hon. Arthur Balfour at Whittinghame, East Lothian. They appeared to be different from *E. gunnii* in that the adult leaves were longer and thinner and the seed capsules had lips on them like *E. urnigera*. They are, however, considered to be a variety of *E. gunnii*, itself a very variable species with some very close relatives such as *E. morrisbyi* and *E. glaucescens*. But Elwes suspected that this Whittinghame variety might be a hybrid between the Cider Gum and the Urn Gum *E. urnigera* which share the same habitat.

A specimen at Whittinghame grew quickly into a remarkable tree standing 19.2m × 4m (63ft × 13ft 5in) (at 60cm (2ft)) when Elwes measured it in 1904, with three main stems resulting from frosting in 1861, although it endured a temperature of -19C (-2F) in 1894 without serious injury. Unfortunately it blew down in 1963.

On the whole the Whittinghame variety seems to be hardier than the type; both are really the only eucalypts which will anything like prosper in eastern England.

In the USA the Cider Gum is rated one of the hardiest of the family, and an attractive ornamental.

In New Zealand, especially on South Island, this eucalypt, like so many exotic trees that grow there, reaches dimensions in excess of those achieved in its natural habitat. Trees in a windbreak at Invercargill have a top height of about 30m (100ft).

Golden Rain Tree, Orlando, Florida (p.234)

Silver Lime, Melbury, Dorset (p.235)

Black Gum, Cumberland, West Virginia (p.237)

Cider Gum, Dunstan Hse, Alnwick, Northumberland (p.239)

Pacific Madrone, Vancouver Island, British Columbia (p.243)

Foxglove Tree, Handsworth Park, Birmingham (p.245)

Indian Bean Tree, Oxford University Parks (p.246)

Blue Gum

Eucalyptus globulus

The Blue Gum is a tall evergreen tree up to 60m (200ft) in height, with a fine bole, sometimes clear of branches for 30m (100ft) and more, with rough, persistent bark at the base but peeling in strips elsewhere to reveal a smooth, bluish-grey surface. The slightly ascending branches support a narrow, rounded crown of long, narrow tapering leaves which are dark green and leathery. In young plants the leaves, which are soft, blue-green in colour with a waxy bloom, are heart-shaped and clasp the square stem.

Distribution and habitat

The natural range of the Blue Gum is limited to the south eastern parts of Tasmania and to two very small areas in the extreme south of Victoria – Cape Otway and Wilson's Promontory. It is found from sea level up to 300m (1,000ft) where the climate is cool temperate, without extremes of heat or cold. It fares best on heavy soil on the slopes of moister valleys. On such sites it is occasionally found in pure stands but more usually in mixture with Stringbark *Eucalyptus obliqua*, and with Mountain Ash *E. regnans*. On poorer soils it consorts with Manna Gum *E. viminalis* and, in Tasmania only, with Black Pepper *E. salicifolia* and White Pepper *E. linearis*.

Life history

The large, top-shaped fruit capsules, containing the small, black seeds, are produced in profusion and seedlings often carpet the ground beneath the trees. Their growth is remarkably rapid 1.8 to 2.8m (6 to 9ft) in the first season being by no means unusual. This extraordinary growth continues through the sapling stage and beyond; trees a mere 50 years old can be as much as 45m (150ft) tall. Ultimate heights of 60 to 75m (200 to 250ft) have been attained, with boles clear of branches for two-thirds of the height and girthing over 6m (20ft). Elwes says that growth when reaching maturity becomes very slow and the tallest of the trees may be as much as 300 years old. However, a sawmiller in Australia reported that he never cut up a Blue Gum older than 75 years.

Timber

The light yellow-brown timber is heavy, hard and durable. It is used in both light and heavy construction work, being highly regarded for purposes where bending is required. It is also used for poles, sleepers, mining timber and pulpwood.

Introduction, performance and use as an exotic

The Blue Gum was discovered by Labillardière in Tasmania in 1799. Mitchell says that it was introduced into Britain in 1829, and it was certainly brought to Europe (Naples) as early as 1832. The oldest tree known to Elwes was one at Tresco, Scilly Islands, which was planted in 1850. It never reached more than 12m (40ft) in height, as its top was blown off, but it was 2.7m (9ft) in girth when it was blown down in 1891. The largest specimen mentioned by Elwes was at Rozel Bay, Jersey. It was planted in 1862, was 33.5m × 3m (110ft × 10ft 3in) in 1892 and produced seed from which seedlings were raised. It was killed in the severe frost of the winter 1894-95. Elwes gives photographs of two trees: one at Kilmorie, Torquay, which was planted in 1897 and was 16m × 1.2m (53ft × 4ft) when the photo was taken in 1910, and the other at Penmere, Falmouth. It was planted around 1865 and had grown to 29m × 2.3m (95ft × 7ft 9in) by 1911, with a 12m (40ft) bole clear of branches.

It is doubtful whether any of these trees or any of the many others mentioned by Elwes are still alive today; for even if they survived severe frosts or, even more lethal to them, persistent cold winds, they would probably have been blown down by gales, as their rapid growth outpaces the support they receive from their roots. However, the Blue Gum sprouts vigorously from stools even after apparently being killed by frost. A tree in the Kyles of Bute, 'killed' by 13 degrees of frost in 1895, sprouted again from the base and was 16.4m × 1.1m (54ft × 3ft 9in) in 1905.

More recent measurements include the following:-

Abersoch, Gwynedd, 26m × 2.5m (85ft × 8ft 6in) 1986
Laxey Green, Isle of Man, 36m × 3.9m (118ft × 12ft 10in) 1978
Powerscourt, Co. Wicklow, 38m × 3.3m (125ft × 11ft) 1985
St Macnissi School, Co. Antrim, 31m × 7.6m (102ft × 25ft) 1982

Apart from the area round Falmouth in Cornwall this tree does not survive for long in Britain, but in Ireland it seems to do much better. In that country there are some very large specimens which have grown astonishingly quickly. At Avondale, Co. Wicklow, a 31-year-old tree was 32m (105ft) tall; and there is a tree at Glengarriff, Co. Cork, which measures 41.7m × 5.9m (137ft × 19ft 7in) and is only about 50 years old.

The Blue Gum is extensively planted in the warmer parts of the USA, especially in California where it was introduced in 1856, although most of the oldest trees now growing in that State probably result from seed sent to a Mr Nolan, an Oakland nurseryman, in 1861. In many parts of the States, the Blue Gum can be seen in shelterbelts and screens as well as in the streets; and it is also grown in plantations for pulp. There too it grows very rapidly and reaches heights of 40m (130ft) or more in as little as 30 years.

This eucalypt is by far the most commonly planted member of the genus worldwide, for shelter, shade and for screens, as well as in plantations, especially in India and South Africa. In the Nilgiri Hills of south India a tree in a 90-year-old plantation measured 70m × 5.7m (232ft × 19ft).

Pacific Madrone

Arbutus menziesii

The Pacific Madrone occurs widely in the understorey of the great forests of the Pacific Coast of north west America and can grow to a height of 24m (80ft). It is an evergreen with big, glossy, leathery leaves which are paler green and sometimes nearly white beneath. The round-topped crown is formed by the upswept, sinuous branches whose thin, shiny, reddish-brown bark peels off in thin scales. Lower down the main trunk of old trees the bark is purplish and finely cracked into narrow ridges, but in young trees the often multiple trunks are smooth, shiny and red down to the ground. (See illustration facing p.241)

Distribution and habitat

The northern limit of the Madrone's natural range is at the Seymour Narrows on the east coast of Vancouver Island and on the opposite coast on the mainland of British Columbia. The range stretches south between the coast and the Cascade Range in Washington and Oregon, down as far as the coastal ranges of Southern California.

This tree grows under exceedingly variable climatic and site conditions; in areas where the annual rainfall is as much as 4,216mm (166in) or as little as 457mm (18in), and on deep alluvial soil sites near the coast as well as on dry, thin, rocky soil at altitudes of 1,830m (6,000ft). But it reaches its maximum development in the Redwood forests of North California, where it can be up to 24m (80ft) tall. On poor soils at high altitudes it is often no more than a large, spreading shrub.

As an understorey tree of the coastal forests the Madrone is found growing beneath Redwoods *Sequoia sempervirens*, Douglas Fir *Pseudotsuga menziesii*, Western Hemlock *Tsuga heterophylla* and Lawson Cypress *Chamaecyparis lawsoniana*, whilst on the mountain slopes it consorts with, and may even be dominant with, Sugar Pine *Pinus lambertiana*, Ponderosa Pine *P. ponderosa*, Digger Pine *P. coulteri* and Canyon Live Oak *Quercus chrysolepis*.

Life history

The flowers of the Madrone are small, white and bell-shaped, rather like Lily-of-the-Valley, hanging in clusters. From them are developed bright red, 1.3cm (0.5in) berry-like fruits containing up to six angled seeds. Large crops are produced fairly regularly but even on good moist, well-drained soil seedlings are never abundant, and root suckers are the major source of reproduction, resulting in the trees being commonly found in clumps.

On good sites, heights of up to 24.4m (80ft) are common, with considerable girths. The recorded girth of one specimen was over 8m (27ft) at 1.3m (4.5ft) from the ground.

Timber

The timber of the Madrone is fine textured, hard, heavy and reddish brown in colour. It is used for small cabinet work, turnery and fuel and formerly for shuttles in weaving machines.

Introduction, performance and use as an exotic

The Madrone was discovered by Archibald Menzies in 1792 when he called it (quite correctly) 'the oriental strawberry tree' and added 'its peculiar smooth bark of reddish brown colour will at all times attract the notice of the most superficial observer'. But it was left to David Douglas to introduce it to Britain in 1827. It was soon found to be less hardy than the European species, *A. unedo*, but Loudon reported, in 1838, that it grew quicker, and that plants were available at 7s. 6d (38p) each. Clinton Baker had a tree at Bayfordbury, Herts,. which was 9m (30ft) tall at the age of 25 years. Elwes, writing in 1906, recorded the largest he had seen to be a specimen at Bassetwood, near Southampton which measured 15.2m × 0.9m (50ft × 3ft 2in). Gardner measured the same tree in 1960 and found it to be 18.3m × 1.6m (60ft × 5ft 5in).

A number of fine specimens are to be found, mainly in the south of England, measuring between 18 and 24m (60 and 80ft) tall, but the tree is generally not hardy enough to thrive in Scotland or the north of England.

Surprisingly for an ericaceous plant, the Madrone will grow on calcareous soils. Amongst recent records are the following trees:-

Borde Hill, Sussex, 19m × 1.3m (62ft × 4ft 7in) 1989
Coverwood, Surrey, 25m × 2.6m (82ft × 8ft 10in) 1986
Fota, Co. Cork, 19m × 2.4m (62ft × 8ft 3in) 1987
Hergest Croft, Herefords., 25m × 2.1m (82ft × 7ft 2in) 1985
Killerton, Devon, 19m × 1.8m (62ft × 6ft 1in) 1987
Savannah, Bagshot, 24.4m × 2.1m (80ft × 7ft 1in) 1982

In the USA, the native Madrone is rated as one of the most handsome broadleaved evergreens, but being difficult to transplant and in the habit of dropping leaves, fruit and/or bark throughout the year, it is not planted as often as it deserves. It does not thrive in the eastern states.

Foxglove Tree

Paulownia tomentosa

The Foxglove Tree is one of the most beautiful of the temperate flowering trees from China. The open-domed crown of thick, sinuous branches and hairy twigs bears the distinctive, hairy leaves which are heart-shaped at the base, pointed at the top with two shallow, pointed lobes, and varying in size from 12 to 25cm (5 to 10in) long. The bole, often branching low down, is clad in greyish-brown, mottled bark, marked with pale, longitudinal fissures. The beautiful, mauve, foxglove-shaped flowers are borne, 10 to 15 at a time, on hairy stalks 30cm (1ft) long. (See illustration facing p.241)

Distribution and habitat

This tree is a native of mountainous and upland areas of central China, often growing as high up as 1,220m (4,000ft). It is a minor component of the forests on the foothills bordering the Great Plains along the east coast of China, in Kiangsu and Anhwei Provinces. It also occurs amongst other species in what appear to be naturally regenerated forests which are again growing up in once cultivated areas of the Plain itself. Here it is associated with such trees as Tree of Heaven *Ailanthus altissima*, Walnut *Juglans regia* and Chinese Thuya *Thuya orientalis*.

Life history

The flower buds which are formed on the flower stalks during the summer, remain exposed on the tree all winter long, to open in May before the leaves. The fruits are brown, two-celled, pointed capsules 1 to 5cm (0.5 to 2in) long. They open in the autumn to scatter a mass of tiny membraneous winged seeds. Growth is very rapid, but the tree never attains great sizes; 15m (50ft) in height and 1.8m (6ft) in girth being about average.

Timber

The timber is very light in weight and colour, and is easily worked, producing a satiny surface. In ancient China it was used for making lutes. In Japan it is particularly favoured for making wardrobes and chests-of-drawers because of its tendency to absorb moisture and prevent the forming of mould during hot, damp weather.

Introduction, performance and use as an exotic

The Foxglove Tree, named after Anna Paulownia, daughter of Czar Paul I, was introduced into Britain by seeds from China in 1838. Some were sown in the Horticultural Society's Chiswick Garden, but many of the original trees from these perished in the severe winter of 1860.

The largest tree which Elwes knew of was at Westonbirt, Glos.; possibly the same one which, according to Mitchell, measured 26.8m × 2.2m (88ft × 7ft 6in) in 1988. All the dozen or so trees which Elwes measured were in Southern England or East Anglia and they varied between 7.6 and 17m (25 and 56ft) in height. One at Ickleton, Cambs., regularly produced flowers and seed from which plants were raised. Another, at Whitbourn, Worcs., was raised from seed gathered in the Vatican Garden in Rome in 1888. In 1902 it was 7m × 1.2m (23ft × 4ft) with a crown spread of 7.6m (25ft).

Recent measurements include the following trees:-

Corsham Court, Wilts., (planted, 1940), 15.2m × 1.8m (50ft × 6ft 3in) 1989
Glasnevin Botanic Gardens, 19.8m × 1.5m (65ft × 5ft 5in) 1989
Mottisfont Abbey, Hants., 15.8m × 1.9m (52ft × 6ft 9in) 1988
Weatherden Hall, Kent, (planted, 1951), 15.8m × 2.1m (52ft × 7ft) 1988
Winchester Cathedral Close, 14m × 1.6m (46ft × 5ft 8in) 1988

Only in the south of England does the Foxglove Tree flower with any regularity, and even then rarely producing seed, not only because warm sunshine is necessary for the buds to form but also because these buds stay exposed on the tree all winter and are frequently killed. However, this tree can be cultivated almost anywhere as a foliage plant. If the stem of a sapling is cut off at ground level in the spring, quick growing shoots will appear bearing extra large versions of the attractive leaves.

On the continent of Europe the Foxglove Tree is much planted as an ornamental. In France and Italy it flowers profusely and sets seed.

In the USA, it is extensively planted in the eastern states where it is greatly prized for its spectacular mauve flowers which are produced by trees grown as far north as New York. It also thrives in the Pacific States where, in some areas it has become naturalised on waste ground.

In Australia, it is planted for shade along roadsides and as an ornamental, particularly in coastal areas where it stands up well to winds.

Indian Bean Tree

Catalpa bignonioides

The Indian Bean Tree or Common Catalpa comes from central and eastern North America, usually with a short trunk. The light brown bark is tinged with red and separates on the surface into large, thin irregular scales. Large, heavy brittle branches form a low, wide-spreading crown. The thin, light green leaves are grouped in threes; they are big, very attractive and heavily veined. The purple, crinkly-edged flowers, grouped on 25cm (10in) panicles, are a spectacular sight in May and June. They give way to thin, dark brown pods, 45cm (18in) long, which dangle from the branches, like bootlaces, all winter. (See illustration facing p.241)

Distribution and habitat

So many Indian Bean Trees have for so long been planted over such a wide area of the USA that the exact natural distribution is unclear, but it is probably south west Georgia, west Florida, central Alabama and Mississippi; and naturalised throughout the southern Atlantic states as well as in Kentucky and Tennessee. In the supposedly natural range the tree is most common on the banks of rivers, but it is never found in pure stands.

Life history

The flowers appear in May or June and the fruits ripen in the autumn; but the pods, containing the 2.5cm (1in) long, winged, silvery seeds, do not open until the following spring when they split along their length to release their contents.

The seedlings grow vigorously, eventually reaching some 18m (60ft) in height and 3m (10ft) in girth. The tree is not long-lived.

Timber

Although the timber is coarse-grained, soft and far from strong, it is durable in contact with the soil and is used almost exclusively for fencing posts.

Introduction, performance and use as an exotic

The Indian Bean Tree was introduced into Britain in 1726 by Mark Catesby who first described it. Loudon lists numerous specimens in southern England which, in 1838, were over 12m (40ft) tall, the tallest being at Syon House, Middlesex, which topped 15.8m (52ft). But Elwes found only its dead stump in 1906 when he recorded two others at Wilton, Wilts., and Arundel, Sussex, both of which just exceeded 15m (50ft) in height. This seems to be about the limit of height growth in present day specimens, 9m (30ft) being more usual. The fact that none of Loudon's trees and few if any of Elwes' are still with us suggests that this tree is not at all long-lived in Britain. Neither is it really hardy in northern England or Scotland, for the fat, fleshy twigs ripen late and are often cut back by early autumn frosts. It really does best in southern areas where hot summers occur, and in towns and cities where high temperatures are maintained throughout the 24 hours. In such places this tree is a very popular ornamental; in leaf handsome, despite late flushing, in flower magnificent and, after leaf fall, interesting with the long, black pods hanging in bunches on thick, spreading branches.

Recent measurements include the following trees:-

Bath Botanic Gardens, 12.2m × 2.7m (40ft × 9ft 5in) 1989
Deans Court, Dorset, 19m × 2.1m (62ft × 7ft 3in) 1986
Royal Botanic Gardens, Kew, (Bell Lawn), 12.2m × 2.7m (40ft × 9ft 4in); (Temple House), 11m × 2.4m (36ft × 8ft) 1981
St. James Park, London, 15.2m × 2.4m (50ft × 8ft 2in) 1981

The Indian Bean Tree is extensively planted as a street tree in southern Europe.

The cultivar 'Aurea', the leaves of which remain a rich yellow throughout the season, is sometimes seen in gardens and parks.

The native Common or Southern Catalpa is planted in the USA as an ornamental, but the type tree is not so popular as the so-called Umbrella Catalpa, *C.b.* 'Nana', usually seen as a graft high on the understock of the species.

Another species, the Western Catalpa *C. speciosa*, with a more limited range in the central eastern states, is a bigger tree, up to 36m (120ft) tall. It has never been much planted in Britain, where it was introduced as late as 1880 with high hopes of its becoming a timber tree. But, of the 150 plants raised from seed sent by Sargent in that year, not one could be found by Elwes in 1910. A plantation of this catalpa was set up near Marlborough by the Marquess of Ailsbury and the trees were 1 to 2m (3 to 6ft) tall in 1907, but they came to nothing because they persistently failed to ripen their wood early enough in the autumn.

Despite its popularity in the USA, where it is bigger, better and more handsome than the common species, the Western Catalpa is rarely planted in Britain, possibly due to its very slow growth and its reluctance to flower as freely as the Indian Bean Tree.

Bibliography

ANDERSON, Mark L. *The Selection of Tree Species*, Oliver and Boyd, Edinburgh, 1950. CANADIAN DEPARTMENT OF FORESTRY, *Native Trees of Canada*. Bulletin 61, 6th. Ed., Queen's Printer and Controller of Stationery, Ottawa, 1961. COATS, Alice M., *Quest for Plants*, Studio Vista, London, 1969. BEAN, W.J., *Trees and Shrubs Hardy in the British Isles*, 7th. Ed., John Murray, London, 1950. COLLINGWOOD, G.H. & BRUSH, W.D., *Knowing Your Trees*, American Forestry Association, Washington D.C., 1949. DAVIES, John, *Douglas of the Forest*, Paul Harris Publishing, Edinburgh, 1979. ELWES, H.J. & HENRY, A.H., *The Trees of Great Britain and Ireland*, Privately Printed, 7 Vols., Edinburgh, 1906-1913. EVERETT, T.H., *Living Trees of the World*, Thames & Hudson, London, 1969. FORESTRY COMMISSION, *Exotic Forest Trees in Great Britain*, Bulletin No. 30, H.M. Stationery Office, London, 1957. FORESTRY AND TIMBER BUREAU, *Forest Trees of Australia*, Government Printing Office, Canberra, 1957. FORTUNE, Robert, *Three Years Wandering in Northern Provinces of China*, John Murray, London, 1847. FORTUNE, Robert, *Residence Among the Chinese*, 1853 to 1856, John Murray, London, 1857. GILBERT-CARTER, H., *British Trees and Shrubs*, Clarendon Press, Oxford, 1936. HADFIELD, Miles, *British Trees*, Dent, London, 1957. HADFIELD, Miles, *Landscape with Trees*, Country Life, London, 1967. HARRIS, Esmond & Jeanette, *Field Guide to the Trees and Shrubs of Britain*, Reader's Digest Association, Ltd., London, 1981. HORA, Bayard, (Editor), *The Oxford Encyclopedia of Trees of the World*, Oxford Univ. Press, 1981. HUI-LIN LI, *The Origin and Cultivation of Shade and Ornamental Trees*, Univ. of Pennsylvania Press, Philadelphia, 1963. JACKSON, A. Bruce, *The Identification of Conifers*, Edward Arnold, London, 1946. JAPANESE FOREST TECHNICAL ASSOCIATION, *Illustrated Important Forest Trees of Japan*, Chikyu Shuppan Co. Ltd., Tokyo, 1964. JAMES, N.D.G., *The Trees of Bicton*, Blackwells, Oxford, 1969. JOHNSON, Hugh, *The International Book of Trees*, Mitchell Beazley, London, 1973. LEATHART, Scott, *Trees of the World*, Hamlyn, London, 1977. LITTLE, E.J., *Sixty Trees from Foreign Lands*, U.S. Agricultural Handbook No. 212, U.S. Dept. of Agriculture, Washington, D.C., 1961. LOUDON, J.C., *Arboretum et Fruticetum Britannicum*, Longman, Orme, Brown, Green and Longmans, London, 1838. MITCHELL, Alan, *Conifers in the British Isles*, Forestry Commission Booklet No. 33, H.M. Stationery Office, London, 1972. MITCHELL, Alan, *A Field Guide to the Trees of Britain and Northern Europe*, Collins, London, 1974. MIROV, N.T. & HASBROUCK, J., *The Story of Pines*, Indiana Univ. Press, Bloomington, 1976. PEARSE, S.A., *Ornamental Trees*, Collingridge, London, 1961. PIM, SHEILA, *The Wood and the Trees* (a biography of Augustine Henry), Macdonald, London, 1966. PRESTON, J.P., *North American Trees*, 3rd. Ed., The M.I.T. Press, Cambridge, Massachusetts, 1976. ROSS, C.R., *Trees to Know in Oregon*, Extension Bulletin 697, Oregon State Univ., Corvallis, 1978. ROYAL FORESTRY SOCIETY, *Quarterly Journal of Forestry*, Vols. 1-76, 1906 to 1982. SARGENT, C.S., *Manual of the Trees of North America*, Dover Publications, New York, 1949. SEYMOUR, William, Botanist Explorers of Two Continents. *History Today*, Aug. 1979, Pp. 497-505. STAINTON, J.D.A., *Forests of Nepal*, Hafner Publishing Co., New York, 1972. STREETS, R.J., *Exotic Forest Trees in the British Commonwealth*, Clarendon Press, Oxford, 1962. TROUP, R.S., *The Silviculture of Indian Trees*, Revised Ed., Controller of Publications, Delhi, 1975. U.S. DEPT OF AGRICULTURE, *Silvics of Forest Trees of the United States*, Agricultural Handbook No. 271, U.S. Govt. Printing Office, Washington, D.C., 1965. VEDEL, H. & LANGE, J., *Trees and Bushes*, Methuen, London, 1960. VEITCH, John, *A Manual of Conifers*, Royal Exotic Nursery, Chelsea, London, 1881. VERTREES, J.D., *Japanese Maples*, Timber Press, Forest Grove, Oregon, 1978. WANG CHI-WU, *The Forests of China*, Harvard Univ., 1961. WILKINSON, Gerald, *A History of British Trees*, Hutchinson, London, 1981. WYMAN, Donald, *Trees for American Gardens*, Macmillan Publishing Co., New York, 1979.

Index